INTERNATIONAL ACCLAIM FOR

The Weather Makers

"Simple, doable advice, superb science writing and fascinating speculation combine to make *The Weather Makers* a great book, one even skeptics will like."

—*Edmonton Journal*

"This is one of the most important books of this young century. It is an urgent call-to-action that we cannot afford to ignore."

—*David Suzuki*

"The finest account of the overwhelming science behind global warming. Flannery gives us a terrifying glimpse of the future."

—*Robert F. Kennedy, Jr.*

"This authoritative and maddeningly important book will fuel dinner arguments, spark school debates and rudely challenge the self-satisfied truffle-eaters and climate deniers among Calgary's oil elite."

—*The Globe and Mail*

"A straightforward and powerfully written look at the connection between climate change and global warming. It's destined to become required reading."

—*Publishers Weekly* (Starred Review)

"It would be hard to imagine a better or more important book."

—Bill Bryson, author of
A Short History of Nearly Everything

"At last here is a clear and readable account of one of the most important but controversial issues facing everyone in the world today. If you are not already addicted to Tim Flannery's writing, discover him now."

—Jared Diamond, author of
Guns, Germs, and Steel

"If you are not yet convinced of the gravity of the [climate change] problem, or our capacity to solve it, you should buy and read this compelling book."

—*The Age* (Australia)

"Essential reading."
—Ronald Wright, author of *A Short History of Progress*

"All who read *The Weather Makers* will be left wiser and able to appreciate how fragile our climate is and how it is this generation who must act to protect it."

—Tony Blair,
Prime Minister of the United Kingdom

"*The Weather Makers* is proof that truth isn't just stranger than fiction, it is also far more terrifying. Flannery has done a superb job of collating and explaining the huge volume of scientific evidence and laying out what is inevitable, and what can and can't be changed."

—*Hamilton Spectator*

"With *The Weather Makers*, Tim Flannery delivers an almighty wallop. . . . The general reader can absorb it and feel enlightened; the scientific reader can, and must, use it as a springboard for further research."

—*The Monthly* (Australia)

"*The Weather Makers* shows us that we no longer have any excuse for letting our governments deceive us about the earth's vulnerability. . . . Sobering."

—*The Vancouver Sun*

"Of the doomsday clocks ticking toward midnight, climate change is the most fearful. Understanding is the first step toward salvation."

—John Polanyi, Nobel Laureate

"Finally, a book about a global crisis that people can understand. All of us who are dubious, or skeptical, or can't make sense of the passionate warnings about climate change will find in this book a clear distillation of the salient facts and their meaning."

—Sharon Butala, author of
The Perfection of the Morning

"Tremendously informative."

—Joy Williams, author of *Ill Nature*

"Presented with a vast array of information in a readable and convincing way, *The Weather Makers* shows clearly that decisive action is needed now."

—Chief Emeka Anyaoku, President,
World Wildlife Fund International

"This is the book the world has been waiting for—and needing—for decades."

—Professor Peter Singer,
internationally renowned author and ethicist

"Flannery weaves the science, politics and the economics together in a tale as frightening as it must have been when Hitler was marching across Europe."

—John Passacantando,
Executive Director, Greenpeace USA

"An urgent call for action. . . . A powerful and persuasive book, sure to provoke strong reaction."

—*Kirkus Reviews*

NOW or NEVER

NOW OR NEVER

*Why We Need to Act Now
to Achieve a Sustainable Future*

TIM FLANNERY

HarperCollins*PublishersLtd*

Published by HarperCollins Publishers Ltd.

First Canadian Edition

HarperCollins Publishers Ltd
2 Bloor Street East, 20th Floor
Toronto, Ontario, Canada
M4W 1A8

www.harpercollins.ca

Flannery, Tim F. (Tim Fridtjof), 1956–
Now or never : why we need to act now to achieve a sustainable
future / Tim Flannery.

ISBN 978-1-55468-604-9

1. Sustainability. 2. Climate and civilization. 3. Climatic changes.
4. Climatic changes–Government policy. I. Title.

QC903.F52 2009   304.2'5   C2009-903022-5

Printed in Canada
DWF 9 8 7 6 5 4 3 2 1

CONTENTS

CONTENTS

RESPONSES

REPLY

FOREWORD
by David Suzuki

Human beings are an infant species, appearing in the last 0.001 percent of the time that life has existed on Earth. For most of the 150,000 years of our species' existence, we were hunter-gatherers, carrying all our possessions in a constant search for food and materials. Even after the agricultural revolution ten millennia ago, we lived within limited confines in the company of a few dozen people. We have been local, tribal animals for almost all our time on Earth.

But suddenly we have become a geological force, altering the physical, chemical, and biological makeup of the planet as no other species has ever done. We have embraced the benefits of our newly acquired powers with little regard for the consequences within the biosphere. But now we have to ask, "What is the collective impact of all 6.7 billion human beings?" — and it is very difficult to assess. Even when we do consider how we are all affecting our surroundings, we find there are no mechanisms to respond as a single species for our own benefit.

It has long been my contention that at the time of our emergence as a species on the plains of Africa, we gave no hint of our explosive development into a dominant force in only 150 millennia. That's because our evolutionary advantage was hidden in our skulls. The human brain conferred an enormous memory, insatiable curiosity, and impressive creativity that more than compensated for our lack of physical and sensory capacities. Accumulating knowledge through experience and imagination, we invented the notion of a future; and in so doing, we found we could influence that future. Using our knowledge and memory, we could look ahead, anticipate dangers and opportunities, and thus deliberately choose to take actions that avoided the dangers and exploited the opportunities. Foresight was our great advantage and was a key part of our enormous success as we spread across the planet.

Today, we are the most numerous mammal on Earth, and our huge ecological footprint (that is, the amount of land and water needed to meet our demands) has been amplified beyond that of any other species by our technological muscle power, voracious appetite, and global economy. It has only been forty-seven years since Rachel Carson told of the costs of our technological prowess in her influential book *Silent Spring*. Despite her prescient warnings, pesticides are used today in far greater amounts and many are far more toxic than those used in 1962.

Our capacity to look ahead has been greatly amplified today, with scientists, supercomputers, and telecommunications; and ever since *Silent Spring*, the warnings of scientists have become more urgent. But now we are turning our backs on the very way that so successfully got us to our current position of dominance.

In 1988, the environment was the number one concern of people around the world. That year, Prime Minister Margaret Thatcher of the United Kingdom declared, "I'm a greenie," and George H. W. Bush promised, if elected, to be "an environmental president." In 1988, Brian Mulroney was reelected prime minister in Canada and, to show he cared about the environment, he appointed his brightest star, Lucien Bouchard, as minister of the environment. I interviewed Bouchard three months later and asked what he felt was the most urgent environmental issue for Canadians. His instant response was, "Global warming." When I asked how serious it was, he replied, "It threatens the survival of our species. We have to act now."

That year, 300 scientists met in Toronto to discuss the atmosphere. They were convinced there was evidence that global warming was occurring and that people were causing it. In a press release, they declared that global warming represented a threat to human survival second only to nuclear war, and they called for a 20 percent reduction in greenhouse gas emissions below 1988 levels in fifteen years. Scientists had spoken, the public was concerned, and politicians had gotten the message. Had we acted accordingly, we would be far beyond the Kyoto target and well on our way to the deep reductions we now know we have to make.

But we didn't respond by taking on the challenge. Politicians didn't have the stomach to take the criticism for spending big bucks to reduce greenhouse gas emissions when they wouldn't even be around to take credit for it fifteen years later. Many environmentalists, including me, felt it was a "slow-motion catastrophe" and there was time to focus on more urgent issues like clear-cut logging. But most egregiously, corporations began to spend millions on a campaign to confuse the public, calling climate change "junk science," supporting articles and Web sites to dispute the evidence, and funding a few "skeptics" to spread disinformation. And it worked. (See *Climate Cover-Up* by James Hoggan.)

Tim Flannery's *The Weather Makers* was a wake-up call. A best-selling book, it made the impact of climate change real and personal and, like Al Gore's *An Inconvenient Truth*, moved a wide audience to take the issue seriously. But as countries moved with glacial reluctance to make big reductions in their greenhouse gas emissions, glaciers themselves were melting with unprecedented speed. The most authoritative voice on climate, the Intergovernmental Panel on Climate Change, has issued updates that have become more and more urgent, even as scientists announce unanticipated rapidity of change.

Flannery's latest message, *Now or Never*, is that we have passed the tipping point for climate change and are approaching a point of no return where we will not be able to do anything about it except hang on for the final ride through very turbulent times. For too long, we have pulled our punches to avoid being dismissed as sensationalists, alarmists, or extremists, even when the science warranted extreme statements. We have urged individual actions like changing lightbulbs and turning off computers while economies, energy use, and emissions continued to rise.

The scientific foresight that enables us to look ahead now demands that we take the gloves off and tell it like it is. We are heading toward a precipice at breakneck speed and we have to slow down and, very soon, turn onto a different road. If we fail to act with the urgency Flannery demands, then our foresight poses a terrifying fate.

Can we make the kinds of major shifts that Flannery suggests climate change demands? Of course. If we don't, we will be left in a far more precarious state, as changes that we can't even anticipate assault us. When Japan attacked Pearl Harbor on December 7, 1941, the American Pacific fleet was severely damaged. Americans didn't roll over and seek peace or decry the cost of all-out war. Americans had no choice—they had to make every effort to win. That's one way: to let matters develop and deal with the consequences when they crop up. The scale of response should mimic a war effort—but there's a better way.

I was beginning my last year in college in 1957 when the world was electrified by the announcement, on October 4, that the Soviet Union had launched Sputnik, a basketball-size satellite, into orbit. In the ensuing months, U.S. Army, Air Force, and Navy rockets all blew up on the launch pad as the Soviets announced the first animal in space—the dog Laika. Then Yuri Gagarin became the first man in space, a team of cosmonauts was launched, and Valentina Tereshkova was the first woman cosmonaut. It was a frightening time as the Russians' advantage in science and engineering confronted the Americans' failures.

Americans didn't give up because the Soviet Union was too far ahead or because the cost of competing would ruin the American economy. Instead, the United States began to spend billions on research, universities, and students. It was a glorious time—I was then a foreign student in the United States—and there were grants and jobs widely available. When John F. Kennedy announced the plan to land Americans on the moon, the public rejoiced at the challenge. And once it had beaten the world to the moon,

the United States reaped completely unexpected benefits from its investment—cell phones, round-the-clock
television news, and GPS. In 2007, half a century later,
every Nobel Prize in science went to an American
lab—all because in 1957, Americans decided to make
an all-out effort to confront the Soviet juggernaut.
That's been the American way, and that is what is
needed now to confront the most serious of all challenges—climate change.

DR. DAVID SUZUKI, cofounder of the David Suzuki
Foundation, is an award-winning scientist, environmentalist, and broadcaster, as well as a world leader in
sustainable ecology. He is the recipient of UNESCO's
Kalinga Prize for Science, the United Nations Environment Program Medal, and the Global 500. He is a
fellow of the American Association for the Advance-
ment of Science.

NOW or NEVER

In the Year Four Billion

We succeeded in taking that picture, and if you look at it, you see a dot. That's here. That's home. That's us. On it, everyone you ever heard of, every human being who ever lived, lived out their lives. The aggregate of all our joys and sufferings, thousands of confident religions, ideologies and economic doctrines, every hunter and forager, every hero and coward, every creator and destroyer of civilizations, every king and peasant, every young couple in love, every hopeful child, every mother and father, every inventor and explorer, every teacher of morals, every corrupt politician, every superstar, every supreme leader, every saint and sinner in the history of our species, lived there on a mote of dust, suspended in a sunbeam.

—Carl Sagan, 11 May 1996

The image that moved Carl Sagan to such poetic magnificence was taken by Voyager 1 on 14 February 1990. The vessel was 4 billion miles from home—a mile for every year of Earth's existence—when it captured that image, and in it Earth is nothing more than a minute

1

blue dot, all but lost in the immensity of the cosmos. At the time Sagan described our home so beautifully he had just half a trip around the sun—six months— to live; and he well knew that the "mote of dust" that had carried him on his life journey is an extraordinary place, for it is the only living planet we know of in all the vastness of the universe.

With the twenty-first century nearly a decade old, Sagan's description resonates more powerfully than ever. Our despoliation of Earth's life-support systems seems to mark us as the destroyer of our own civilizations; and as the planetary crisis we have created deepens, it is certain that no savior will ride across the cosmos to rescue us from ourselves. There is no real debate about how serious our predicament is: all plausible projections indicate that over the next forty to ninety years humanity will exceed—in all probability by about 100 percent—the capacity of Earth to supply our needs, thereby greatly exacerbating the risk of widespread starvation, or of being overwhelmed by our own pollution. The most credible estimates indicate that we are already exceeding Earth's capacity to support our species (this is called its biocapacity) by about 25 percent. With global food security at an all-time low, and greenhouse gases so choking our atmosphere as to threaten a global climatic catastrophe, the signs of what may come are all around us.

Everyone knows what the solution is: we must begin to live sustainably. But what does that actually mean? "Sustainability" is a word that can mean almost anything to anyone. Whether used by cosmetics advertisers or fruit sellers, it is bandied about as if it were the essence of virtue. Yet so recent is the word that my spell-checker doesn't recognize it.

Wikipedia, which is increasingly taken as a fount of all knowledge, defines sustainability as "a characteristic of a process or state that can be maintained at a certain level indefinitely." This is hardly a moral definition this, or indeed—in light of the second law of thermodynamics—a feasible one. Many environmentalists opt for a more practical meaning: "living in such a way as not to detract from the potential quality of life of future generations." And here we find a definition in harmony with a commonly voiced aspiration: to "try to leave the world a better place than we found it." This essay is in part an inquiry into the causes of our common failure to realize this heartfelt

3

desire—even though it is held by almost every individual on Earth.

If we accept the environmentalists' definition, living sustainably does not involve any particular morality beyond an extension of the Eighth Commandment: Thou shalt not steal—even from future generations. A society that limited itself to such a narrow aspiration, however, could be a barbarous place. As in the movie *Soylent Green*, why waste a corpse? Why worry about the distribution of wealth? Any meaningful inquiry into sustainability must surely be broader than this, and thus be as much a philosophical and moral discussion as a scientific one; for sustainability pertains to us—our innate needs and desires—as much as it does to the workings and capacities of our planet. A real search for sustainability involves a broad vision—indeed, it encompasses many flash-point issues: Is eating meat appropriate in a sustainable world, for example? And what of animal rights—and human rights—and religion, and democracy, and the free market, and war? Although a detailed consideration of how these issues can be squared with a fully sustainable future is beyond the scope of this essay, such questions will continually arise as we examine clear, practical solutions to our most urgent problems.

Where does science fit into this inquiry? In human affairs there is often a great difference between aspiration and achievement. Even a society that has developed a moral and philosophical framework ideally suited to attaining a sustainable future may fail to accomplish that if it lacks knowledge of how the world works, and of how its own practices and technology are affecting Earth's life-support systems. Accurate scientific knowledge of Earth and its processes is vital to the pursuit of sustainability. And so I begin this investigation with two questions, which, even if they cannot be definitively answered, can nevertheless guide us in our search: What is our purpose as a species? And how does Earth work?

The wellsprings from which we derive meaning in our lives are intensely personal. My own search for meaning has led me to the belief that this generation— the generation living in the early twenty-first century— is destined to achieve an extraordinary transformation, unique in the 4-billion-year history of Earth, and that this transformation will influence the fate of life from now on. Geologists talk of the dawning of a new

5

geological period called the Anthropocene, which is characterized by pervasive human influence on the Earth's processes. But perhaps the Anthropocene age will truly have dawned only when humanity uses its intelligence to help regulate those processes for the good of life as a whole.

The great complexity and order created by evolution through natural selection have led to the concept of Gaia: Earth as a self-regulating, evolving system. James Lovelock, the originator of the Gaia hypothesis, illustrated it by showing how Earth as a whole maintains the temperature of the planet's surface within bounds that are conducive to life, recycles nutrients, and regulates the chemistry of the atmosphere and oceans to the same effect. In short, living things absorb carbon and heavy metals from the atmosphere and oceans, while at the same time producing oxygen. Life is such a powerful force that this activity keeps the atmosphere and oceans out of chemical balance with the rocks in a way that keeps Earth habitable. The Gaia hypothesis is a way of describing how our living planet works as a whole.

We have long understood — from biblical teachings and practical experience — that we are naught but earth: ashes to ashes, dust to dust, as the English burial service puts it. Indeed, "Dust thou art, and unto dust shalt thou return" (Genesis 3:19) are among the oldest written words that have come down to us. Yet although we have long understood that we are earth, it is equally true, but almost never said, that we are Earth (i.e., part of planet Earth) as well. We are Earth by virtue of the fact that every one of us has been shaped by the process of evolution through natural selection: the process that led to the exceedingly complex and highly ordered structures of life and its ecosystems. And this fact has a profound implication: Earth was not made for us; rather, we were made for this Earth.

This implication about our purpose goes against some of the most powerful currents in western civilization, including the Christian tradition I grew up in. In fact, it is diametrically opposed to such currents, for it asserts that we have evolved to serve Earth, and that our great distinguishing characteristic—our intelligence—is not ours alone, but Gaia's as well, for it is destined to be used by Gaia for her own purposes. James Lovelock took the term Gaia from the ancient Greeks: it was the name of their earth goddess. I believe that over the course of the twenty-first century we will again come to serve our Earth goddess, and perhaps even revere her.

Looking at the current condition of Earth, you might be tempted to see humanity as an enemy of Gaia, but to do so would be a mistake. We are obviously part of Gaia, and, just as obviously, as animals in the Gaian system we must kill (even if we kill only vegetable matter) in order to survive. Gaia is all about the giving, taking, and reprocessing of life. Conceiving of ourselves as outside of and antagonistic to Gaia is, I believe, a terrible mistake, for it leads us to consider actions necessary for our survival as somehow wrong. As animals we must eat, and eating implies taking life. Striving for a bloodless, painless world of perfect morality and zero impact on nature is delusional. Even more important, it blinds us to what I believe is the true purpose, according to the Gaian perspective, of our existence.

I believe that the deepest significance of the twenty-first century can be glimpsed in the hierarchical structure of life on Earth. Here lies the potential for sustainability and the transformation of our existence. Guided by evolution, the history of life has been one of increasing complexity and increasing efficiency.

The eminent evolutionary biologist Stephen Jay Gould argued that life has not increased in overall complexity, because simple life-forms such as bacteria still constitute by far most of life on Earth. Yet from a Gaian perspective, this theory overlooks the undeniable spread and increasingly sophisticated development of life. Life has spread from its origins on the bottom of shallow seas 3.5 billion years ago to almost all parts of Earth's rind. Some 540 million years ago, creatures learned to burrow into the sediments of the seafloor. Then they colonized land, the air, and the ocean depths. Furthermore, as life evolved and spread, it has improved. The reproductive systems and the use of energy in many evolutionary lineages have become more efficient over time; and in these lineages the brain (the command-and-control system) has grown larger and more sophisticated relative to the brains of ancient ancestors.

Large, highly evolved creatures such as mammals play a disproportionately important role in influencing the carbon cycle and other ecosystem processes. There is no doubt that their evolution has increased Gaia's ability to control planetary life-support systems, for as mammalian metabolism has become more complex and efficient, so has that of the planet as a whole. Six hundred million years ago, when there was little or no complex life on Earth, thermostatic control was so poor that the planet repeatedly froze right to the equator, an event known as "snowball Earth." Since the rise of complex life, such events have not recurred.

Evolution through natural selection is a blind process that takes place only by means of variation (within populations) and failure to reproduce (of the less well adapted). That's why Richard Dawkins likened the process to a "blind watchmaker." But now, after 4 billion years, the evolutionary process has arrived at a potentially powerful and swiftly responsive command-and-control system that may serve Gaia as a whole. That system is our own human intelligence and self-awareness. It is my belief that we humans are poised to become, from now on, the means by which Gaia will regulate at least some of its essential processes.

Is it right to say that we are Gaia's self-awareness? Gaia's brain? I believe it is. After all, we commonly talk about our own self-awareness, yet rarely question whether our toes, for example, are aware of the beautiful starry night that our brain is taking in. Admittedly, the human body is far more highly integrated than are

Gaia's disparate parts. But it is undeniable that we are a part of the Gaian whole. Whether there is a Gaian meaning to our existence or not, acknowledging that we are an influential part of Gaia requires a change in the way we interact with Earth's life-support processes. After all, the brain does not despoil the body that it is part of, for to do so would be to destroy itself. Admittedly, the brain is expensive to run. Our own brain, which constitutes just 2 percent of our body by weight, greedily takes about 20 percent of all the energy we consume. As Gaia's intelligence, humanity will doubtless impose a heavy tax on Gaia, yet this burden cannot be so great as to bankrupt the system that supports it.

Gaia's potential for intelligent control is very recent: it arose abruptly toward the end of the twentieth century, after humans had plumbed the depths of the oceans, revealed Earth's internal structure and history, and photographed it from deep space. Scientists such as Carl Sagan were the first to recognize the full significance of these achievements, yet because we have not focused on sustainability, even today the great mass of humanity is unaware of their true import.

By the twenty-first century the achievements of pioneers such as Sagan had opened the way to a limited

11

understanding of how Earth works. Here, scientists such as James Lovelock led the way, and as a result of their efforts we can now describe in some detail how Earth recycles minerals and nutrients, how atmospheric and oceanic chemistry is maintained, how the surface temperature of our planet is regulated, and how biodiversity is protected from external shocks. It is as if, by the late twentieth century, we finally lifted the hood of our planetary vehicle and saw the sophisticated engine concealed within. Then, at the dawn of the new century, we began to understand how it worked.

Such deep understanding of Earth's self-regulatory systems is invariably empowering. Just as surgery could not progress without Harvey's discovery of the circulation of the blood, humanity could not hope to positively influence Earth's thermostat without knowledge of the carbon cycle. If the twentieth century was the century of technological triumph, then the twenty-first century may be an even more significant moment in planetary history: the century when our knowledge of Earth's processes must be put to use. Within the lifetimes of many people reading this essay, after 4 billion years of self-regulation, Gaia will pass from an unconscious to a conscious means of control. Either that or we will fail to achieve sustainability, and Gaia's newly attained consciousness—which is made possible only by our global civilization—will vanish, perhaps to be lost forever.

It is all too possible that we will fail to achieve sustainability, and that the blind watchmaker will once again—through variation of organisms and through the failure of ill-adapted organisms to reproduce—reset the balance of a severely diminished living Earth. Well before we were ready to assume control of the planet, humans were already influencing Earth's processes in ways that threatened to end in global catastrophe. Acting without an awareness of the consequences of our actions, or even a sense of responsibility, we were (from the perspective of Gaia's purpose) immature. Now our fate and that of our planet will be determined by the rate at which we, as a species, can mature and develop a new sense of responsibility. I fear that if we are to avoid catastrophic failure, we will need to learn very fast: learn, indeed, on the job. Our search for sustainability is thus an uncertain experiment, which must inevitably see setbacks and failures. Succeeding at it in the long run will be the greatest challenge our species has ever faced.

THE CLIMATE PROBLEM

There was a time, about 100,000 years ago, when there were just 10,000 people on Earth. A century ago there was 1.5 billion of us, and now there are 6.6 billion. It is estimated that just forty years from now there will be 9 billion. With luck and good management, our population will not grow beyond this point. But some estimates see the number swelling by 1 billion or more in the century after that. That's 10 billion people, on a planet that once held 10,000. Such a burden of human flesh, which all needs to be housed, clothed, and fed, will exacerbate all our environmental woes. Yet who can we ask to get off? The truth is that if we wish to act morally, we can influence population numbers only slowly. So, although it's important to focus on decreasing the population as a long-term solution, we cannot look to it as a solution to the immediate crises.

One problem facing humanity is now so urgent that, unless it is resolved in the next two decades, it will destroy our global civilization: the climate crisis. The warming trend is real and accelerating, and our pollution is responsible for it. All but the most ignorant, biased, and skeptical now admit this truth, and it's underlined by the findings of the Intergovernmental Panel on Climate Change (IPCC). This body of world experts is painfully conservative, for the members work by consensus and include government representatives from the United States, China, and Saudi Arabia, whose assent is required for every word of every finding. In its *Fourth Assessment Report* (which was published in November 2007), the IPCC blandly stated that the warming trend was "very likely"caused by humans (that means the cause is at least 90 percent certain), and subsequent research has confirmed this, dismissing the idea that sunspots or any other cause proposed by the skeptics could play a role.

The farther into the climate system we try to follow the consequences, the less certain the link with human activity becomes, yet even here great advances are being made. In its *Fourth Assessment Report*, the IPCC thought it only "likely" (66–90 percent certain) that there was a relationship between the warming caused by humans and various changes in Earth's physical and biological systems. In May 2008, however, the largest and most definitive study yet on this subject was published in the world's leading science journal, *Nature*. It announced a clear link between a huge number of changes in the natural world and human-caused warming. The researchers' database included such diverse observations as changes in polar bears' behavior, stream flow, the timing of grape harvests, the flowering time of plants, and bird migration. This study is a landmark in our understanding of just how profoundly we are influencing the very Earth processes that give us life.

As we seek to understand our increasing impact on our planet, it's helpful to think about how we are shuffling matter among the three great organs of Gaia and thereby creating an imbalance. Gaia, the living planet Earth, is like a tree in that it is not alive all the way through. Instead, life is restricted to a thin "rind" that extends seven or eight miles below Earth's surface and about fifteen miles above it. This rind is composed of three great organs: the Earth's crust, air, and water. We must consider how matter flows through these three organs, how they interact, and how life in turn influences them, if we are to understand our planetary home.

Earth's crust may seem passive—a mere substrate on which life exists—but it is deeply influenced by the presence of life. Today, the energy captured by plants through photosynthesis contributes three times more energy to Earth's overall geochemical cycles (the weathering, burial, and formation of rocks) than geologic activity such as the formation of mountains and volcanism, and in the past this extra energy played an important role in forming the Earth we know. During its first 600 million years (before life arose), Earth had no continents. A remarkable recent study suggests that the extra energy captured by algal and bacterial life led to the development of Earth's continental crust.

How so? All of Earth's original crust was formed from the dark volcanic rock known as basalt, and even today basaltic crust underlies the oceans. Continental crust is formed by the weathering of basalt. Weathering processes separate the lighter elements in the basalt (particularly the silica-rich elements) from the denser ones. These lighter elements, once they have been compressed and heated, form granite, and give rise to the continental crust. Scientists postulate that the vast amount of basalt weathering required to form the continents could have occurred only if algal and bacterial life captured huge amounts of solar energy, some of which was used to manufacture chemicals, such as oxygen and acids, that helped break down the rocks to form sediment. This finding is still debated, but the inference is that without the contributions of early life to the geologic cycle, there would never have been earth beneath our feet.

Earth's crust is dynamic, and its dynamism is particularly vital to life. The continents shift on large "plates," so that every 300 million years or so the plates bearing the continents coalesce, creating an Earth with a single large continent surrounded by sea. Then the plates break apart again, only to come together in another cycle. No one understands precisely what makes Earth's plates move, but the force of gravity, circulation within the molten mantle of the Earth, and the pull of the moon are all thought to exert an influence.

With regard to life, the most important thing about this movement of the plates is its effect on the recycling of minerals and salts. Where plates collide, the rock underlying one continent is thrust under another and is melted. As a result, mountain ranges and volcanoes are formed, and rivers erode the mineral-rich rocks, creating fresh new soil. This renewal, along with the slow grinding of glaciers, fertilizes life on Earth with the minerals that are essential to plant and animal growth.

Of all the minerals recycled through Earth's crust, carbon is the most critical for this discussion. On planets without life, such as Mars and Venus, the great bulk of the atmosphere is made up of CO_2. On our living planet, in contrast, CO_2 is just a few parts per 10,000 of the atmosphere. The reason for the difference is that over the aeons, enormous quantities of carbon have been drawn into Earth's crust, where today they remain in the form of coal, oil, natural gas and limestone.

If the movement of the plates is important to life on land, it is absolutely vital to life in the sea. The waters of the ocean are recycled, through evaporation and precipitation, through Earth's rivers every 30,000 to 40,000 years, and with each recycling, rivers leach salt from the rocks, which is carried into the sea. You might deduce from this that the oceans are growing saltier. In the nineteenth century this is exactly what scientists thought. Assuming that the oceans were fresh upon their formation, and knowing the rate at which salt was carried into the oceans by rivers, they estimated Earth to be just a few tens of millions of years old, and then coupled this incorrect finding with a belief that a sort of salty doomsday awaited us a few million years hence, when the oceans would have become as salty as the Dead Sea.

The truth is far more remarkable. Earth's oceans have maintained a relatively steady level of saltiness for billions of years, and they do so thanks to the mid-ocean ridges, where Earth's plates are pulled apart, allowing the ocean basins to grow. As the oceanic crust pulls apart, magma comes to the surface and the ocean penetrates this new, hot rock. Hydrothermal vents form, and through these eventually all of the ocean water in the world circulates. It takes 10 million to 100 million years for all the water in the oceans to pass through the hydrothermal vents, but as it does so the chemical structure of the seawater is altered by the extreme heat, and salt is removed. This recycling of the oceans through evaporation, rainfall, and rivers every 40,000 years, and through the movement of the crust at the mid-ocean ridges every 10 million to 100 million years, keeps the saltiness of the sea constant. It is a remarkable thought that all this is made possible by the continents and their movement—continents that life itself may have helped create.

Earth is the water planet, and water, in its three states—vapor, liquid, and solid—defines and sustains Earth. The principal part of its liquid state forms the second organ of Gaia—the oceans, which cover 71 percent of Earth's surface. Solid water, mostly in the form of glacial ice, covers a further 10.4 percent. Water is essential to life because the various electrochemical processes that constitute humans and other life-forms can occur only within it. The ocean was almost certainly the cradle of life, and it remains life's most expansive habitat. The volume of the oceans—about 330 million cubic miles—is eleven times larger than all the land above the sea. And whereas land is populated by life only at its surface, the entire volume of the oceans is capable of sustaining life.

The oceans are the most important means by which carbon is drawn from the atmosphere. Indeed, when considered on a timescale of centuries, historically they have been the only carbon sink that counts. And today, with more carbon in the air, these sinks have much more to absorb. Some of the carbon absorbed by the ocean is used by algae, and some remains dissolved in the water, where it forms carbolic acid. Some of the carbon taken in by algae falls to the ocean floor when the algae die and sink, and there it is destined to form carbonate rock, thereby removing the carbon more or less permanently from the atmosphere. The carbolic acid that remains in the water, however, is very different. As it builds up, it causes the ocean to acidify; and acidity damages life, including the algae that sequester the carbon. Ocean acidification is a much more urgent threat than we previously thought, and it is most advanced in the north Pacific Ocean.

The north Pacific Ocean is so full of life that it seems like a fantasyland. When I first encountered it, walking along the shore at Tofino near Vancouver in British Columbia, I was awestruck by the drifts of mussel and oyster shells almost as long as my foot, the gigantic barnacles and other oversize sea wrack. Offshore, gray whales abounded within a few yards of the beach, as did seals and killer whales. For me, coming from a dry and impoverished land, the sheer abundance of life — and titanic life at that — was almost beyond my reckoning.

The unique fecundity of the north Pacific is caused by the same factors that render it exquisitely vulnerable to acidification. The great frozen continent of Antarctica sits at the center of Earth's oceanic system, for much of the deep and intermediate ocean water is exported from its icy fringe. This icy origin dictates that the average temperature of the ocean is a mere 38 degrees Fahrenheit, which is a good thing indeed for life, as frigid water is full of dissolved oxygen and so can support life in the oceans from bottom to top. There is, however, one important exception to this: the north Pacific, which, because of its unique configuration, is the only ocean not cooled and oxygenated by Antarctic waters.

Instead, deep water, depleted of oxygen and rich in CO_2 (and thus acid), wells up here, bringing with it the nutrients that feed the region's oversize life. The result is a fecund ocean, but one where the depth at which organisms can lay down calcareous skeletons is perilously close to the surface. Thus, anything that requires a shell or skeleton has difficulty surviving at depth in the north Pacific. In other oceans, living things can lay down skeletons to a depth of 5,000–8,500 feet, but in the north Pacific they cannot do so below 400–1,800 feet. This is why stony corals, which are found in every other ocean, are absent from the north Pacific. Increasing CO_2 in the atmosphere has already caused a rise in the boundary below which life cannot lay down a skeleton in the north Pacific, to 100–325 feet. Scientists are now warning that in just a few decades, creatures living in the far north Pacific may be unable to lay down skeletons even at the surface. And this would mean an end to all those oysters, mussels, crabs, and lobsters that this fecund ocean yields. Indeed, ultimately it will probably mean an end to the whales and seabirds as well, for without krill, what will they feed upon? And in time, if the problem persists, all the world's oceans will suffer the same fate.

The atmosphere is the smallest, most vulnerable, yet most vital of Earth's organs. To look up into the blue vault of the heavens in an effort to judge its size or importance is profoundly misleading, for the atmosphere appears to stretch on endlessly. Actually, the atmosphere is a gossamer-thin wrapping, insufficient even to swathe Earth's tallest peaks in breathable air. To get an idea of its actual size, we need to carry out a thought experiment. Imagine compressing the gases of the atmosphere about one-thousandfold—until they become a liquid (this is necessary for a valid comparison). Then imagine comparing the volume of this liquid with that of Earth's oceans. If you could do that, and could see the result, you would discover that the aerial "ocean" is just one five-hundredth the size of Earth's great water oceans. The size of a pollution sink is a prime indicator of its vulnerability. We all know that a small creek or lake is far more likely than a larger one to be damaged by a given volume of pollution—say, sewage. Because the oceans are 500 times larger than the atmosphere, their pollution history has been dramatically different. As we shall soon see, this simple fact will dominate human considerations in the twenty-first century—at least during its first half.

Our shuffling of matter between Gaia's three great organs—crust, air, and water—is at the heart of the problem of climate change. The problem results in large part from digging up the dead—vast amounts of fossilized, once living matter in the form of coal, oil, and natural gas—and burning it. This liberates the ancient carbon that was once in living things, and allows it to reside again in the atmosphere and oceans. Another source of carbon is the destruction of forests and degradation of soils. Since the beginning of the industrial revolution it has added about 200 billion tons of carbon to the atmosphere. The carbon imbalance we have created in these ways is enormous: in just 200 years, the proportion of CO_2 in the atmosphere has risen by around 30 percent—from 2.8 parts per 10,000 to 3.8 parts per 10,000 by 2008. And it is growing. In 2008 the annual carbon emissions of humanity reached 10 billion tons, an amount that caused the concentration of atmospheric CO_2 to increase by 2.2 percent. Not for 55 million years has such an imbalance existed.

A New Dark Age?

In 2006 James Lovelock published a book that bluntly laid before us the consequences of the carbon imbalance. *The Revenge of Gaia* was published in its author's eighty-seventh year, and it is as bleak and penetrating a perspective on human folly in regard to the environment as has ever been written. Lovelock argues that Gaia's climate system is far more sensitive to greenhouse-gas pollution than we imagine, and that the system is already trapped in a vicious circle of positive feedback. "It is almost as if we had lit a fire to keep warm," Lovelock opines, "and failed to notice, as we piled on fuel, that the fire was out of control and the furniture had ignited." Although there is still time to avert a catastrophe, Lovelock believes that humans lack the foresight, wisdom, and political energy required to do so. Instead, he predicts, before the twenty-first century is out our global civilization will have collapsed and a new dark age will have descended on us. Only a few survivors (perhaps just one out of every ten alive today) will cling to the few remaining habit-able regions, such as Greenland and the Antarctic Peninsula.

The events likely to destroy our civilization include dramatic rises in sea level, which will flood coastal cities and some of the best agricultural land; changes in rainfall; extreme weather; and the disappearance of the glaciers that act as dams and whose meltwaters provide our most productive agricultural regions with water in the growing season. The ensuing starvation, warfare, and chaos will be the greatest scourge, for in Lovelock's projected dark age the warlords will be armed with nuclear weapons.

How probable is it that this bleak vision will come to pass? Because of new scientific data and technological analysis we are better placed than ever before to determine the scale of the threat and its imminence. Let's begin with a new analysis of work done by the Intergovernmental Panel on Climate Change (IPCC) in 2001. In its *Third Assessment Report*, the IPCC published a series of projections concerning key indicators of Earth's climate system. These included estimates of how swiftly Earth's average temperatures might increase over the course of the twenty-first century, how much the oceans would rise, and how quickly CO_2 would accumulate in the atmosphere. The projections had an upper and a lower limit, and they encompassed quite a wide range of possibilities. The projection concerning temperature, for example, indicated that the increase might be as little as 2.5 degrees Fahrenheit, or as much as 10.4 degrees. From the perspective of human survival, the difference between 2.5 degrees and 10.4 is profound. Humanity can probably cope with a warming of less than 3 degrees, but a 10.4-degree warming would be truly catastrophic, heralding an ice-free world, and most likely human tragedy on the scale envisaged by Lovelock.

At the time these projections were published, skeptics described them as unbelievable and grossly inflated, and widely proclaimed in the popular press that they amounted to scientific scaremongering. By 2007, however, scientists had five to six years' worth of real-world data under their belts, allowing them to revisit the projections to determine their accuracy, at least over the near-term, early portion of the curve. What they discovered should have been reported the front page of every newspaper on the planet. Astonishingly, in every instance the real-world changes were at the upper limit, or worse than even the worst-case scenario presented by the IPCC. The full implications of these new studies have yet to sink in among those negotiating the global treaty that is supposed to protect humanity from dangerous climate change. The negotiators continue to argue on the basis of the old projections, which call for action far less urgent than what is actually required. Worse, the negotiations grind on as if we had an eternity to achieve outcomes. Lovelock, who seemed like just another prophet of doom just two years ago, appears to have been right after all—unless, that is, we can rouse ourselves to take immediate action.

From mid-2007 onward I've found it increasingly difficult to read the scientific findings on climate change without despairing. Perhaps the most dispiriting changes are occurring at the north pole. The sea ice that covers the Arctic Ocean is an ancient feature of our planet. It has glistened brightly into space for at least 3 million years, and over that time a host of organisms, from plankton to walruses and narwals, have adapted to life on and under it. But its importance to Gaia is far greater than as a home for an unusual fauna: the northern ice acts as a refrigerator that cools the entire planet. It does this by reflecting the sun's energy away from Earth. During the summer, the sun's rays beat down upon it twenty-four hours a day, but because the ice is bright, 90 percent of that energy (which averages 22 watts per square foot) is deflected back into space. Where the ice is absent, however, the dark ocean is revealed, and it soaks up all that solar energy and turns it into heat.

Around 1975, scientists noticed that the Arctic ice had begun to melt away. At first the rate was hardly worrying, and indeed many thought that it might just be part of a long-term cycle. But the trend continued, so that by 2005 the Arctic ice cap had been melting at a rate of around 8 percent per decade for thirty years. At that rate, it would have taken until 2100 or thereabouts for the ice cap to disappear altogether, and to many people, that was a comfortably distant date. But then, in the summer of 2005, a dramatic change occurred. The rate of melting accelerated, so that about four times as much ice melted, compared with previous summers. As at the onset of the melting trend, scientists were hoping that this was a freak or cyclic event, and that in a subsequent summer the melting would once again slow. But the summer of 2006 saw almost as much ice lost as in 2005. Then, during the summer of 2007, the very worst loss of Arctic ice ever witnessed occurred.

These changes in the Arctic have left many scientists worried that the region is already in the grip of an irreversible transition. During the winter months, the Arctic is now warming four times faster than the global average, and the existing temperature increase year-round already exceeds 3.5 degrees Fahrenheit. As a result, profound shifts are occurring in species distribution: some fish stocks in the Bering Sea, for example, have already moved by 500 miles. None of the models used to predict how the Arctic will alter as it warms has been able to replicate any of these changes. None, indeed, is even remotely accurate, so as we try to predict the region's future, we are truly flying blind.

The extent of uncertainty prevailing among scientists is illustrated by a straw poll conducted among experts on the Arctic in March 2008. They were asked whether they thought that summer 2008 would see a regrowth of the Arctic ice. The winter had been a cold one, and the great loss of ice the previous summer had been exceptional, leading the majority to say that a regrowth of the ice cap was likely. Yet by May 2008 the melting had begun once more, and the average daily loss of Arctic sea ice was, on average, 2,300 square miles per week greater than for the same period of 2007. By June the losses had become so severe that one Norwegian expert was saying that 2008 might see the Arctic's first ice-free summer. As it happened, 2008 saw a slight improvement in the extent of the ice (about 115,000 square miles) over the previous year. But the following winter brought extremely poor ice formation, with the area of ice the same as the all-time record low of 2007. It's now clear that the Arctic's first ice-free summer can be no more than a few decades away, and indeed may come to pass in just a few years.

What will happen during that first iceless summer? Most likely, not much at all, for it will take several summers' worth of energy to warm the surface of the Arctic sea to a point where dangerous changes are generated farther south. If recent history is anything to go by, during that first iceless summer the skeptics will say, "See, we told you that there was nothing to fear from an ice-free Arctic," and those who don't know any better will grasp at the reassurance. But each year thereafter, the ocean at the top of the world will inexorably warm, and the temperature gradient that

controls climatic zones across the northern hemisphere will shift. It's difficult to know precisely how
that will affect humanity, but if we look back at the
last time in Earth's history such a great warming
occurred — 55 million years ago — we see an omi-
nously different world. Back then, lemurs prolifer-
ated in the rain forests of Greenland, and the tropics
were covered by a spiny, thin, alien-looking vegetation, which is today extinct. No one knows how
quickly the world's climate altered then, but one
cannot help fearing what a similar change might
mean for humanity today.

So swift are the changes already occurring in the
Arctic that much of the human response to the crisis thus far has been rendered hopelessly inadequate.
The warming, for example, has accelerated the rate
of melting of the Greenland ice cap, which is now van-
ishing by between sixty and seventy cubic miles per
year. Public policy responses and political discourse,
meanwhile, are based on a previous rate of loss of just
twelve cubic miles per year. And this melting really
does have immediate relevance, for the Greenland ice
cap sits on land, and as it melts, it contributes to a rise
in sea level. Even the most committed conservation-

ists have been forced to rethink their strategy. Neil Hamilton, director of the WWF International Arctic Programme, said in May 2008, "We [WWF] are no longer trying to protect the Arctic," because it is too late. He believes that the region's first ice-free summer may arrive before 2013, and admits that he has no idea what the Arctic might look like in 2050.

New ramifications of rapid warming are continually being discovered. In 2006 scientists realized that the sea can die as a result of severe global warming. Indeed, it has died, several times during Earth's history, and when it dies, it takes most life on land with it. Evidence of a dying sea comes from the sediments laid down on its floor. At different times enormous deposits of black shale have formed, and these are the source of much of Earth's oil. Oil, of course, is derived from living things, and it can form only when the organic matter that gives rise to it doesn't rot. Very little, if any, oil is forming in the oceans today because their depths are so filled with oxygen that living things can exist there; and life in the abyss, as life always does, efficiently uses and recycles whatever organic matter rains down on it. It therefore takes a dead ocean—or at least one whose depths are dead—to

35

make oil, and oceans begin to die when the abyss is starved of oxygen.

The ocean circulation is vigorous today because the poles are cold and the equator is warm. The source of most of the deep ocean water is around the Antarctic; that is why ocean water is so cool, about 34–35 degrees Fahrenheit. This cold water (which can hold lots of oxygen) permits life in the depths. The cold poles and warm equator also cause the winds that drive surface currents, and the resultant surface mixing helps oxygenate the waters.

The most devastating example of oceanic death occurred 250 million years ago, when 95 percent of all life perished. Just what occurred then is only now beginning to be understood, largely because of a breakthrough in geochemistry. It was realized that living things such as bacteria, which rarely leave conventional fossils, nevertheless leave a chemical signature of their existence in rocks. Geologists studying rocks in Western Australia that dated to the Permian-Triassic extinction of 250 million years ago discovered traces of the unique lipids (fatty molecules) made by strange kinds of bacteria known as purple bacteria and green sulfur bacteria. These bacteria thrive only in waters that are well lit by the sun, yet are low in oxygen and high in hydrogen sulfide. Today, such conditions exist only in very restricted and unusual environments, such as the "jellyfish lakes" of Palau. Yet the story preserved in the rocks reveals that most if not all of Earth's oceans resembled this environment 250 million years ago.

The steps leading to the death of the oceans have been reconstructed as follows. First, a sudden increase of CO_2 and methane in the atmosphere causes rapid warming of air and sea, which disrupts ocean currents and warms the depths. Increased warming of the poles brings winds and surface currents nearly to a standstill; and because of slowed circulation, and the fact that warm water holds less oxygen than cold water does, the ocean depths become deprived of oxygen. In this environment, bacteria that don't require oxygen multiply, and they emit huge volumes of sulfur. Eventually, the sulfurous, oxygen-starved water reaches the sunlit zone, and then the green sulfur bacteria flourish, producing huge volumes of toxic hydrogen sulfide, which enters the atmosphere in great belched bubbles, destroying much life on land. The gas rises high into the atmosphere, where it destroys the ozone layer, and the increased ultraviolet (UV) radiation devastates what is left of life on Earth.

What does an Earth with a dead ocean look like? Peter Ward, a palaeontologist and expert in his field, imagines it as follows:

Look out on the surface of the great sea itself, and as far as the eye can see there is a mirrored flatness, an ocean without whitecaps. Yet that is not the biggest surprise. From shore to the horizon, there is but an unending purple colour—a vast, flat, oily purple, not looking at all like water. . . . The colour comes from a vast concentration of purple bacteria. . . . At last there is motion on the sea, yet it is not life, but antilife. Not far from the fetid shore, a large bubble of gas belches from the viscous oil slick-like surface. . . . It is hydrogen sulphide, produced by green sulphur bacteria growing amid their purple cousins. There is one final surprise. We look upward, to the sky. High, vastly high overhead, there are thin clouds, clouds existing far in excess of the highest clouds found on our Earth. They exist in a place that changes the very colour of the sky itself. We are under a pale green sky, and it has the smell of death and poison.

How much time, exactly, do we have to prove Lovelock wrong? In October 2008, Dr. James Hansen (who is arguably the world's leading climate scientist) and eight of his colleagues provided a new, alarming, though still partial, answer to this question. They looked back over the increasingly complete ice-core record, which documents the last 750,000 years of Earth's climatic history, and tried to determine how much warming a given amount of atmospheric CO_2 pollution would produce, and how long it would take to produce this warming. Their most alarming discovery was that, when viewed over the long term, Earth's climate system is about twice as sensitive to CO_2 pollution as the IPCC's century-long projections would indicate. This implies that there is already enough greenhouse-gas pollution in the atmosphere to cause 3.5 degrees Fahrenheit of warming, bringing about conditions not seen on Earth for 2 million to 3 million years, and constituting, according to the authors, "a degree of warming that would surely yield 'dangerous' climate impacts."

Fortunately for us, some—perhaps half—of that warming is currently masked by other pollutants, known collectively as agents of global dimming, which reflect sunlight into space, thus cooling Earth. These include sulfur dioxide (the cause of acid rain), photochemical smog, and tiny particles of carbon called aerosols. All these pollutants are dangerous to human health, and it was for this reason in part that governments in Europe and the United States moved to regulate them long before tackling the greenhouse gases. They are also very short-lived in the atmosphere, lasting only hours to weeks.

Today, China, India, and other rapidly industrializing economies are releasing these agents of global dimming in ever-increasing quantities. Yet because of their effect on visibility and their serious impact on human health, there's good reason to believe that in the near future such nations will move to curb their release. Indeed, in the period leading up to the Beijing Olympics, heroic efforts were being made to do just that over large parts of northeastern China. One particularly effective instrument used to achieve this is a government subsidy for every kilowatt of electricity generated at plants that do not emit sulfur dioxide.

As a result of this scheme, one of China's "big three" providers of electricity—Datang International—had all of its generating plants fitted with sulfur-dioxide scrubbers by the end of 2009, and the competition is not far behind. If no attempt is made to reduce the agents of global warming concurrently with such cleanups of the agents of global dimming, humanity could experience a nearly instantaneous increase in warming that might have catastrophic consequences.

Hansen and his colleagues have arrived at a new understanding of how long it takes for the full warming consequences of a given amount of greenhouse gas to be felt. Two major factors cause a delay in the warming. The first of these, the rate at which the oceans are able to absorb the extra heat trapped in the atmosphere is perhaps the more important, and certainly the more easily determined. According to Hansen, if the delay caused by the oceans alone is considered, then we could expect to feel one-third of any warming caused by a given amount of greenhouse gas in the first few years after the gas is released. Three-quarters of the full warming effect would be felt within 250 years, and all of it within a millennium.

There is a second factor that causes a delay in the warming impact: Earth's ice, which currently covers 10.4 percent of the planet. You can think of ice as a kind of battery that stores cold, and the rate at which ice vanishes from a warming world is a key factor in determining when the full warming impact will be felt. Unfortunately, it is extremely difficult, if not impossible, to predict the decay of Earth's ice fields, mainly because they don't simply melt away like an ice cube. Rather, large portions can collapse spectacularly, spilling into the sea in fragments, where they rapidly melt. Such phenomena cannot be replicated in any of the models used to predict climate change. That is a tragedy, for in the real world the polar and glacial ice caps are altering profoundly and rapidly.

The rapid surface melting of the Greenland ice cap, the collapse of coastal ice shelves that hold back glaciers, a marked speeding of the ice streams that flow through great ice shelves such as the West Antarctic Ice Sheet, and an alarming overall loss of ice are all being observed in the real world, yet we are at a loss to determine how quickly, or how much, they will add to a rising ocean. But one can reasonably speculate. As Hansen and his colleagues put it, "Sea-level changes of several meters per century occur in the palaeoclimate record, in response to forcings slower and weaker than the present human-made forcing. This indicates that the ice may disintegrate and melt faster than previously assumed, and that the warming may be delayed less by the ice than assumed."

In their landmark paper, Hansen and his colleagues make a useful distinction between climatic "tipping points" and "the point of no return." The climatic tipping point is the point at which the greenhouse-gas concentration reaches a level sufficient to cause catastrophic climate change. The point of no return is reached when that concentration of greenhouse gas has been in place sufficiently long to give rise to an irreversible process. Humanity is now between a tipping point and a point of no return, and only the most strenuous efforts on our part are capable of returning us to safe ground. The work of Hansen and his colleagues indicates that we still have a few years before we reach the point of no return, but that there is not a second to waste. This is our greatest challenge, and clearly the path for- ward involves a drastic change in energy use. It also means making full use of the tools we have at our disposal—and inventing new tools—to draw the pollution out of the air and save us from Lovelock's new dark age.

The Coal Conundrum

Hansen and his colleagues summarize the challenge as follows: "If humanity wishes to preserve a planet similar to that on which civilization developed and to which life on Earth is adapted, palaeoclimate evidence and ongoing climate change suggest that CO_2 will need to be reduced from its current 385 ppm to at most 350 ppm." This, they believe, can be achieved only by phasing out all conventional coal burning by 2030, and by aggressively reducing the amount of CO_2 in the atmosphere by capturing it in growing tropical forests and in agricultural soils. That a rapid phaseout of coal is in itself not enough is elegantly illustrated by the fact that the concentration of CO_2 in the atmosphere would remain above 350 ppm for 200 years were a coal phaseout to be achieved within the next decade or two, and nothing else was done. Yet the point of no return is, in all probability, less than twenty years away.

Just how large is the task of replacing the present fossil fuels (in particular conventional coal) with other, nonpolluting fuels? On 3 April 2008, the researchers Roger Pielke, Tom Wigley, and Christopher Green published a study examining the IPCC projections that guide current thinking on the extent to which emissions need to be reduced. Shockingly, they discovered that the IPCC projections underestimate the scale of the task by two-thirds.

The reason for this is that the lion's share of the emissions reductions required in the future are already "built into" the IPCC's scenarios. In other words, the IPCC assumes that these reductions will occur anyway, even in the absence of specific policies aimed at producing the shift. This assumption may seem doubtful, but it was based on the observation that improvements in technologies—particularly, greater efficiency—occur over time. Thus internal combustion engines have become more efficient, as have refrigerators and countless electrical appliances. But can we expect that such improvements will lead to a slowing in the overall rate of greenhouse-gas pollution, regardless of government policy? The answer came when the researchers examined the relevant changes in the real world that had occurred over the first eight years of the twenty-first century.

Dismayingly, they discovered that no "built-in" emissions reductions were occurring: in fact, the efficiency of global energy use (measured as energy intensity) and carbon intensity (pollution) had both risen during this period.

Clearly, the task of combating the climate crisis is far larger than conventional wisdom assumes. In order to establish how much larger, the researchers ran the IPCC projections without assuming any "built in" emissions reductions. They found that the real task is four times larger than the IPCC projections indicate. (And this was assuming that we wish to stabilize atmospheric CO_2 at 500 parts per million rather than Hanson's 350!) To rise to this challenge, humanity will need to implement clean energy technologies approximately ten times faster than is projected by the most ambitious of the IPCC scenarios.

In summary, Pielke and his colleagues note soberly, "The world is on a development and energy path that will bring with it a surge in carbon dioxide emissions—a surge that will only end with a transformation of global energy systems." Indeed, keeping the developing Chinese and Indian economies in mind, they believe that the real surge in CO_2 emissions, if no concerted effort is made, is only at its beginning. This is not to say that humanity is bound to fail. Indeed, during crisis, such as World War II, astonishing breakthroughs in technology and manufacturing have occurred. The problem regarding CO_2 is that, so far, humanity has failed to see the need for urgency. A commentator on this groundbreaking research has pointed out where the real human deficit lies, saying that when it comes to dealing with the climate crisis, "no amount of scenario planning can replace the need for will and leadership."

How abundantly blessed are we with will and leadership? A look at the coal industry and its much-vaunted clean coal initiatives is enough to drive one to despair. For years the U.S. government and industry partners have funded the planning of a pioneering clean coal power plant known as FutureGen. The plant is designed to burn coal with great efficiency, and to capture the resulting CO_2 and store it underground. If the technology proves economical, effective, and safe, it will be a potent tool for combating climate change.

47

The FutureGen project was meant to lead the way toward clean coal, so many were delighted when, in December 2007, after seemingly interminable delays, a site for the plant was finally announced—in Mattoon, Illinois. But then, astonishingly, just a month later, on 29 January 2008, the U.S. Department of Energy announced that it was withdrawing funding from the project. The reasons for this catastrophic decision remain obscure, but lawyers representing Illinois claim that Texas lost the bidding for the plant, and that this cost the project its political support.

The enormous growth in energy generation—mostly coal-fired—in China adds to the urgency of the need for clean coal. Power generation capacity in China is rising from 442,000 megawatts in 2004 to a projected 920,000 by 2010—a doubling in just seven years. That is equivalent to the installation of about 1,300 megawatts of power capacity—the amount generated by a very large power station—every week. In India the situation is only marginally less alarming. India's eleventh five-year plan, which finishes in 2012, calls for the installation of 90,000 megawatts of generation capacity. Much of this will be supplied by burning the country's poor-quality and highly polluting lignite coal. But the expansion of conventional coal-fired power plants in India will also require importing 100 million tons of coal during this period.

When I visited India recently the urgency of national economic development became obvious to me. Four hundred million people there lack access to any electricity whatever, and deep poverty is widespread. It is futile to tell Indians that they should defer development of power plants until cleaner technologies are available, so that we can spare unborn generations climate change. Why, Indians ask, should they penalize people living today for future, uncertain gains, and do this to help solve a problem that is not of their creation?

It is obvious that enormous investment in electricity generation infrastructure will, whether fairly or not, dictate key elements of the world's response to climate change. That's because China and India will not simply knock down their newly constructed power plants in response to the need to reduce emissions. Instead, carbon capture will have to be retrofitted to these plants, and ways found to cover the costs. The bad news is that such retrofitting is even more economically and technologically challenging than building a new clean coal project like FutureGen.

Just how the required technology will be developed, and how such a huge retrofit will be financed, is far from clear. The challenge is all the more difficult because in China the price of electricity is capped. Therefore, power companies cannot pass on rises in the cost of power generation to consumers, and the recent increases in the price of coal are leading to financial losses, so it is not feasible for companies to invest in the new technology themselves. Despite the effect on future investment, the central government is reluctant to raise the price of electricity because inflation, driven by rising food prices, is already straining social harmony. The only practical solution in such a case is for the developed world to help shoulder the cost of reducing the pollution.

One way of achieving that is to allow transfer of funds through a "clean development" mechanism, such as the one available in the European trading scheme, which allows polluters in Europe to pay for emissions abatement in, say, China if that is more cost-effective than reducing pollution themselves. Unfortunately, there are strong signs that in a future carbon trading scheme the United States would allow no such transfers: the Republicans who can prevent passage of bills in the Senate believe that transfers are tantamount to helping the opposition. More fundamentally, as long as carbon capture remains an unproven technology, no funds can be transferred under any scheme. Therefore, there's an urgent need for someone to invest in the development of carbon capture technology.

If the fate of their industry depends on investments in new technology, why, you might ask, are the coal companies waiting for government agencies (such as the U.S. Department of Energy) to foot the bill for clean coal? After all, the price of thermal coal (the kind of coal used in power plants) doubled in 2008 to about $112 per metric ton (from its cost of $56 per metric ton in early 2008). Coking coal (which is used in steelmaking) was doing even better; its price was $300 per metric ton, up from $97 a year before. Prices have since fallen sharply, but with such windfall profits accruing to the industry, surely there's plenty of latitude for investment in technologies that offer its only lifeline to the future.

Thus far, investments by coal companies in clean technologies have been insufficient to fund the completion of even a single large demonstration

51

plant. In late 2008, the Swedish energy company Vattenfall did open a 5-megawatt power plant that deploys full carbon capture and storage technology. It is innovative in design, burning the coal in a mixture of exhaust gas and oxygen. But this plant is very small — in fact, it would have to be five times larger just to be included in the European carbon trading scheme — and because of its small scale, the economics of such clean coal remains opaque. Given the years that the coal industry has been first denying this issue and then dragging its feet, we may be tempted to feel that vision, leadership, and will are in short supply in this industry.

Of course, there are reasons for this. Coal mines and coal-fired power plants often have different owners, so while the mines are making a profit, the power generators might be feeling a squeeze. Yet they are ultimately interdependent and all in the same business, and you'd think that the coal industry's association would be busying itself to find a solution. I believe that there is a role for government regulation here. Coal miners should be made to share responsibility for emissions with power companies, for that would focus the industry more sharply on reducing emissions as required. Nobody is actually considering this, although governments are coming to the rescue in other ways. Australia is the world's largest coal exporter and, among the developed countries, depends most heavily on coal. You might think that its coal industry and government would be working hand in hand to solve an urgent problem. Yet in its 2008 budget Australia's government announced a mere $500 million investment in clean coal technology. Billions are in fact required. One option for achieving the required scale of investment would be for Australia to pool its efforts with a reliable partner, such as the United States or Germany, and to impose stiff levies on all coal exports and uses; the revenues would go into a common pot to fund the search for a solution. Nothing like this level of cooperation is occurring, however.

One other aspect of clean coal technology is worth touching on: the reliance on appropriate geologic structures to store CO_2 underground. Among the best of such structures are those that hold natural gas and oil, and where such structures exist near coal-fired power plants, the cost of clean coal will be much reduced. If, however, we envisage replacing every conventional coal-fired plant on Earth with clean coal, the situation looks very different, for the required pipeline infrastructure would be staggeringly large. Indeed, it would probably rival the entire existing pipeline infrastructure deployed by the oil and gas industries combined. With pipeline costs soaring (partly as a result of demand in China), the required pipelines could not, as a practical matter, be in place by 2030. Of course, this argument could be applied to any energy technology that requires rapid ramping up, for all such technologies face severe bottlenecks of one sort or another. I merely note it here to make the point that clean coal technologies can never be a complete, worldwide replacement for existing coal facilities. Globally, renewable energy will have to take a significant portion of conventional coal's market share.

Do not assume from any of this that I believe clean coal technologies to be safe or cost effective. In some circumstances they may prove to be as dangerous as nuclear power and as expensive as solar panels. My point is that the world, and China in particular, has gone so far down the road of using coal as an energy source that we have little choice but to pursue a solution that involves coal.

In April 2008, *Time* magazine included the Australian prime minister, Kevin Rudd, in its list of the world's 100 most influential people. Rudd's significance comes, in part, from his fluency in Mandarin, at a time when China's economic and political influence is on the rise; and of all the issues facing humanity, and China in particular, climate change is the most critical.

One vitally important initiative concerns the transfer of intellectual property involving clean coal to China, by means of a global or bilateral treaty, as swiftly as possible. This is not an entirely novel idea. As long ago as 1919 the patent for aspirin was transferred from the Germans to the Allies at the Treaty of Versailles. My suggestion is that, as a condition of funding the Australian coal industry, the federal government should force it to give any intellectual property in clean coal technology it develops to Chinese power companies for domestic use. Australia's carbon trading scheme, which will be in place by 2011, could also be effective in assisting China to deal with its problems, if the scheme involved a "clean development mechanism" similar to the one operating in Europe.

There are doubtless those in the environment movement who will consider it bitterly unfair that

55

further government subsidies go to the very industry that has lied to the public for decades and has done more than any other to create the climate crisis in the first place. I agree that it's unfair, but I can see no other way of getting out of our present crisis. Perhaps the world would be a better place if environmentalists punished the culprits through the courts, but allowed whatever was required by way of clean coal development in order to give the world the chance at a better future. My own view is that the children and grandchildren of the coal burners will deliver a far harsher punishment than either the courts or our economies are capable of.

America's New Leadership

On January 20, 2009, Barack Obama was sworn in as the forty-fourth president of the United States. His first executive orders dealt with energy security and climate change. His presidency therefore marks a dramatic shift in American policy, and nowhere are the scale and importance of the change more evident than in the American Recovery and Reinvestment Act of 2009. The act, which he signed into law on February 17, commits $40.75 billion to clean energy initiatives, and will stimulate enormous activity in weatherizing housing, energy efficiency, the national electricity grid, public transportation, and cleaner motor vehicles, creating more than 500,000 jobs in the process. The investment in clean energy is enormous, representing an increase of almost 400 percent in U.S. spending. To put this in perspective, total global spending on clean technology in 2008 is estimated to have been $100 billion. Because of its sheer scale, Obama's new investment will have not only national but global significance.

As is widely acknowledged, these encouraging developments are occurring as the world faces an economic crisis the like of which it has not seen for more than seventy years. During 2008, some 69,000 factories closed in southeastern China, and millions have been thrown out of work around the world. The crisis—which is a compound of several economic shocks, including a meltdown in the banking sector and a crisis in American (and increasingly global) housing, and a marked slowdown in almost all sectors of the global economy—defies ready solutions, and it is widely believed that sluggish economic conditions will persist for months or even years. Because of the severity of these problems, and the human misery they could engender, politicians everywhere have been giving them priority. Some environmentalists might feel that the climate crisis has been neglected as a result, but there is good evidence that, in some countries at least, this is not so. Indeed, I believe that the world economic crisis may, in the medium term, help us come to grips with our deteriorating climate. One reason for my optimism is that the economic and climate crises are occurring on different scales. The global treaty to be negotiated in Copenhagen in December 2009 will not even come into effect until 2013, and none but the most bearish are suggesting that the economic crisis will linger that long.

With regard to climate, one of the most positive changes engendered by the economic crisis is a new willingness by both businesses and the public to tolerate government regulation. In fact, government spending and regulation are now seen by many as the only means we have of dealing with the economic meltdown. This new acceptance of government regulation bodes well for climate negotiations, because such acceptance is a precondition for any global treaty. Another positive change is the way leaders are collaborating to deal with the crisis. In 2008 the CEOs of the world's reserve banks met to devise a common approach, and for once it was not the G8 nations that were called on to act, but the G20. This latter organization, which has been in existence only since 1999, is altogether more weighty than the G8. The G8 countries represent 60 percent of global GDP but only 14 percent of the world's population; by contrast, G20 represents nations responsible for 90 percent of the global GDP, 80 percent of the world's trade, and two-thirds of its population. With the help of the global financial crisis, the G20 seems set to take over from the G8 as the forum for discussing cricial issues facing humanity, and because it is far more representative of humanity, it offers a far better chance of brokering a global deal on greenhouse gas emissions.

Despite the increased cooperation and the increased sharing of knowledge, the United States remains far in advance of most countries in explicitly linking solutions for the climate crisis with economic recovery. The economic stimulus packages of many other nations are far less focused, perhaps reflecting the power of special-interest groups such as the housing and energy sectors. As part of its stimulus package, for example, the Australian government sent out checks to citizens just prior to Christmas, in the hope that the money would be spent on consumption during the holiday season. And Australia's 42 billion AUD recovery package, which was passed in February 2009, includes huge giveaways to simulate housing construction, but almost nothing for clean energy (though at the last minute the Australian Greens wrung out

$500 million for clean energy projects). Such policies promote more of the same—more indeed of the consumption that precipitated the climate crisis in the first place.

The United States is only now moving toward mandating a price on carbon pollution, by means of a carbon trading scheme. Europe established its scheme in 2005, and Australia's will begin in 2011. Although such schemes are absolutely necessary to reduce emissions, they alone will not be sufficient to bring about the new industrial revolution needed to save us from a catastrophic climate shift. This point is driven home when one considers the limited influence that the spiraling price of oil and gasoline had on people's choice of transportation during 2007 and 2008. True, a few more smaller cars were sold, but because we don't buy new automobiles often, and because aside from automobiles our options for transportation are so limited, the rising prices did not cause a mass exodus from large cars. To effect a similar rise through implementing a price on carbon, polluters would have to pay $300 per metric ton of carbon, which is far more than anyone is contemplating for a U.S. carbon trading scheme. This suggests that, at least in the case of transportation, there is a need for government regulation of the auto industry that encourages low-emission vehicles—and spending on buses and light rail—alongside carbon trading.

It's not possible, within the compass of this essay, to review all the means of generating electricity without carbon emissions, and their state of development. Instead, I will note one unexpected ray of hope, emanating from Denmark. Electric cars have been "on the horizon" for about a century, and various attempts have been made to commercialize them, all with limited or no success. It's not that electric engines can't do the job—they'll take you from zero to 100 faster than anything else, and they will keep accelerating past 100 for as long as the machine is capable. The problems have been their short range of travel and the difficulty of refueling. These limitations have now been overcome by a remarkable collaboration between two companies: DONG Energy, one of Denmark's wind-energy pioneers, and Better Place; a company dedicated to developing the intellectual property enabling the recharging of electric cars. Throughout late 2007 and early 2008, they worked together on a project to bring the electric car to Denmark.

On 28 March 2008, their plans were announced at a press conference held by the Copenhagen Climate Council. The cars will be fueled from electric plug facilities, and it's hoped that that before the end of 2010 such plugs will be available at one in every six parking spots in Denmark, or from electric-car battery exchanges that will be located throughout the nation.

The system developed by Better Place includes a push-in, pull-out format for all electric cars, which allows batteries to be exchanged as easily and quickly as we currently fuel a car, and a full battery can drive the car 100 miles. Charging takes a little longer, but as it is done at curbside, that isn't much of a problem. Owners of the new electric cars would pay for their electricity much as we pay for mobile phone credit, receiving a monthly bill. In order to encourage the acceptance of electric cars, the Danish deputy prime minister announced that no taxes would be payable on such vehicles until at least 2012. Because taxes on motor vehicles in Denmark are approximately 180 percent of the actual cost of the car, this is a very significant concession. Nor did it hurt that the demonstration vehicle brought to Copenhagen was fast, sexy, and elegant.

63

After the press conference I asked DONG's CEO, Anders Eldrup, why his company intended to spend hundreds of millions of euros on what looked to me like a speculative venture. He explained that the company saw this as an outgrowth of its commitment to wind energy. At present, all the electricity generated by DONG's turbines at night was wasted, since hardly anyone wanted electricity at that time. If the company could find a market for this power in the charging of electric car batteries, that would be a good thing. "We're about creating a virtual oil field in the sky," he said.

Although it will be balm to the soul to see electric cars proliferate in Denmark, the true significance of this initiative is far more profound. What it tells us is that wind energy can compete directly with big oil. Wind is often disregarded as an intermittent source of energy, but when we consider how an electricity grid that includes storage capacity in myriad electric-car batteries would function, this intermittency looks like much less of a problem.

Trees for Security

Reducing emissions is one part of the challenge. Equally pressing is the need to get some of the polluting gas that's already been emitted out of the air. Without corrective action, CO_2 remains in the atmosphere for a century or more once it has been released by burning fossil fuels. As we explore the options, it is worth keeping in mind that we are facing several crises at once: declining oil reserves, ever more perilous food security, and, of course, climate change. If we are to survive the twenty-first century, we must find ways forward that address these issues simultaneously.

For now, high-tech methods of drawing CO_2 out of the atmosphere remain on the drawing board. The strongest possibility of a large-scale drawdown of atmospheric carbon lies instead in changes to global agriculture and forestry. That's because plants represent an astonishingly effective means of carbon capture, each year drawing down approximately 8 percent of the atmospheric carbon. At that rate it would take only twelve years to draw all the carbon out of the atmosphere; but of course plants die and rot, releasing a similar amount of carbon into the atmosphere. And if forests are felled at a faster rate than they can grow, even more carbon is released. If only a small part of the carbon captured by plants could be stored more or less permanently, great inroads into the standing stock of the pollutant could be made.

The best place to begin is where the drawndown carbon is most abundant—the great belt of tropical forest at the equator. Although these rain forests cover only a small proportion of Earth's surface, an estimated two-thirds of all living species live there. The rain forests are also home to hundreds of millions of Earth's poorest and most underprivileged citizens, and to some of its most unsustainable agricultural and forestry practices. We commonly misunderstand how trees grow, imagining that they somehow grow from the soil—from their roots. But this is not the case: trees build themselves from CO_2, which they draw from the air via tiny holes in their leaves called stomata. That is, they grow by withdrawing carbon from the atmosphere and solidifying it into bark, wood, roots, fruit, and leaves. Look at a plant and (if you can imag-

66

ine its roots) you can roughly estimate the amount of carbon it has sequestered over its lifetime.

Tropical forests are prodigious engines of atmospheric sanitation because constant warmth and abundant moisture allow them to grow strongly and continuously all year, so that they sequester carbon more effectively than plants growing elsewhere. Furthermore, such forests play a critical role in stabilizing the climate: they help to keep Earth cool through the transpiration (release) of moisture and the creation of clouds.*

Hitherto we have not valued the service to climatic stability provided by the tropical trees. Instead we have, since time immemorial, hewed and felled them, turning them into exotic timbers such as teak, ebony, and meranti to adorn our homes. Yet for all our ax

---

*Forests growing further north can, under some circumstances, actually contribute to warming. That's because, although they sequester carbon, they can also alter Earth's overall brightness (known as its albedo). This occurs particularly if forests are grown in parts of the northern hemisphere where snow falls duting the winter. If occupied by grass or tundra, such areas are bright for much of the year, and so reflect solar energy back into space. If a dark forest canopy replaces that snow cover, however, then solar energy is trapped and transformed into heat, and this addition of heat can exceed any cooling brought about by the forest's growth.

work, until the nineteenth century the tropical forests had survived pretty well. When Alfred Russel Wallace visited the island of Singapore in 1862, he encountered an old, dense forest filled with tigers, which, as he reported, "kill on average a Chinaman every day." Not long before that—in the seventeenth century—Javan rhinoceroses, tigers, and leopards lurking just beyond the stockade were a danger to the Dutch settlers at Fort Jakarta. At that time, Hong Kong was just a quiet fishing village, and few tropical Asian cities were much more.

It has taken only a century to transform half of the world's tropical forests into farms, cities, and useless, rank grassland. And at current rates of clearing, by 2030 80 percent of the tropical forest that remains will have vanished. The destruction of these forests produces much greenhouse gas; consequently, some developing tropical countries have substantial negative carbon balances. Papua New Guinea, for example, has a greenhouse-gas emissions rate one-third that of Australia, even though Australia is almost ten times as large and burns huge quantities of coal. Globally, 18 percent of all human-caused greenhouse-gas emissions each year result from the ongoing destruction of Gaia's vital rain forests, and since 1800 between 22 percent and 43 percent of all human emissions have come from the destruction of forests.

Traditionally, tropical forests were destroyed as a result of swidden agriculture. This practice involves felling a patch of forest, using the soil for a few years until it is exhausted, then moving on to clear a new patch. At low population densities the practice is sustainable because the forest has a chance to regrow. But as population density increases, the cycle of forest destruction and regrowth shortens until all that can survive is rank tropical grassland growing on impoverished soil. Today, mechanized tropical forestry and agricultural practices greatly amplify the destruction, with the result that the tropics are full of grasslands that are of little use to anyone. If these areas were left alone, eventually the forests would regrow in them; but burning to provide access, and for hunting, prevents this.

Very few people benefit from the ongoing destruction of the tropical forests. Villagers who live in the region lose a renewable source of building materials, food, and medicine, while we in the developed world lose a vital opportunity to stabilize our climate. It is only the loggers, who appropriate the patrimony of the

69

original inhabitants—indeed, the patrimony of all mankind—and turn it into profit for themselves, who benefit, at least in the short term. And that sort of profit taking must surely be regarded as theft, regardless of the letter of the local law.

Is it possible that Earth's rain forests could be saved—even restored—as we work to stabilize our climate? Many schemes to do just this are now under discussion, but all, I feel, suffer from the same fault: they all seek to work through government channels. It's almost a truism that funds given to governments in the tropical, developing world rarely work their way down to the village level. Unless the people who live in and from the forests benefit from their protection, the forests will continue to be destroyed, regardless of what we do or any government does. For this reason, I think that the only way forward is to establish a marketplace linking people like you and me, who are eager to purchase our climatic security, to tropical subsistence farmers who are willing to sell us that security by preserving their forests. EBay shows how such a market might be established, and the growth of Internet-related services shows how rapidly such a scheme could develop. But what is obviously lacking is a means for us to establish direct contact with the farmers.

It is now technically possible to provide Internet access to people in the most remote places on Earth: places without electricity or landline telephones. One way of starting a carbon trading scheme for tropical forests would be to fund a series of trial projects that would provide Internet access to selected primary schools in the tropics. The basic computer training, along with ways to formulate carbon sequestration plans, could be provided by nongovernmental organizations (NGOs), such as WWF or Greenpeace. With their help, villages could post a blog, perhaps accessible via Google Earth, introducing themselves to the world and explaining how they plan to go about protecting existing forest, or to reforest degraded grassland. Then, through an auction system like eBay, those interested in purchasing their climate security could pay for a proportion of the proposed carbon sequestration.

One of the most difficult obstacles facing anyone who wishes to use growing forests to sequester carbon is ensuring the security of the investment. After all, sequestered carbon is unique among crops in that it never leaves the land of the vendor, and so is vulnerable to a change in the vendor's circumstances. Satellite surveillance allows periodic inspection of a carbon investment, and NGOs can undertake first-hand inspections. But an auction system like eBay offers another, more pertinent form of security, for one of the most extraordinary aspects of trading on eBay is the way it enforces honesty between buyers and sellers who have never met, and in all likelihood never will meet, and who live in different jurisdictions. The key is the publication of both the vendor's and the purchaser's reliability records. If either defaults on a deal, this fact is posted and the trader's record is compromised, with the result that other traders will be extremely reluctant to deal with him. Were carbon sequestration in tropical forests to be auctioned on an annual basis, this aspect of the auction system would provide powerful security. Further security could be provided by having an NGO hold the funds in escrow until the carbon sequestration had been verified.

I can imagine two distinct markets developing, both aimed at sequestering carbon in tropical forests. Some corporations and individuals may wish to conserve existing forests along with their carbon and their biodiversity. Appealing species, such as harpy eagles, orangutans, and tree kangaroos, may become emblems of the protected forests, and of the sponsoring corporation. Other individuals and corporations, however, might wish to support reforestation in order to draw CO_2 out of the atmosphere. Because there are vast areas of degraded forest and grassland that could, with little effort and no harm to anyone, be returned to forest cover, this could be a very large-scale market—and one with multiple benefits.

There is no doubt that such a scheme would have to be implemented carefully, with much village education and assistance to prepare communities to enter the global marketplace. Contention over tribal boundaries might need to be setteled; the division of the funds among villagers would have to be agreed on; and how the funds might be spent or invested would need to be discussed. It would be important that the purchasers of carbon not try to dictate the kind of vegetation to be grown. Tropical farmers are shrewd managers of risk—after all, they're never more than a crop failure away from famine. They will doubtless begin by planting things that will benefit them in other ways, such as food trees and trees useful in building. As long

73

as the carbon balance on their land remains positive, we should not be concerned.

As villagers become conversant with the Internet, I can foresee that the computers in the village schools would be used for other purposes. Crops such as coffee might be sold directly to purchasers, rather than to middlemen who take the bulk of the profit. Artifacts and accommodations for visitors stays might be advertised, and eventually pornography would be accessed and western goods would be purchased. Both good and bad things come from joining the global community, yet none of this should prevent us from reaching out to our fellow humans and assisting them to ascend the first rung on the ladder to a better life.

It's my guess that once such a scheme was started, it would be self-perpetuating. In effect, carbon trading could become a way to enter the global community. If the villages initially joining the scheme were seen to benefit, other villages nearby would be likely to band together to buy their own computers and call on the required expertise. If the UN meeting in Copenhagen in December 2009 decides to include forests in carbon trading, this trade could be worth billions per year — enough, perhaps, to lift the world's poorest out of their deep poverty. Of course, even if no such agreement is reached, a voluntary trading scheme (one in which the carbon price is not regulated by means of a government-mandated cap on emissions) could be established. On average, tropical forests draw in 7.4 metric tons of carbon per acre per year, but young, vigorous forests can take in far more. If the carbon price was around twenty dollars per ton, a very substantial trade could be established using just degraded tropical landscapes.

There is a certain natural justice to this solution. The standing stock of greenhouse gas in the atmosphere is approximately 200 gigatonnes (1 gigatonne being 1 billion metric tons), and it has been accumulating since the beginning of the industrial revolution two centuries ago. We affluent westerners, of course, have been the beneficiaries of that revolution, yet the pollution produced as a result will afflict all Earth's people. Indeed, it will have a disproportionate impact upon the poor, because they lack the means to shield themselves against it. It is only just, many argue, that the developed world should repay this "historic debt" in a way that benefits the poorest people on Earth. The extent to which a tropical reforestation program could draw down carbon is not yet clear, but it seems reasonable to assume that within a few decades, about five gigatonnes of carbon could be drawn from the atmosphere each year. This represents 2.5 percent of the historic debt of carbon that has built up in the atmosphere since 1800.

Revolution in the Feedlot

As of early 2008, humanity had just thirty-seven days' worth of grain supply in reserve. It's been said that the ancient Romans had a more generous buffer against famine. As a result of rising food prices, India, the Philippines, and some other nations have now banned the export of certain types of rice. The Haitian government, which acted too late, fell in early 2008 as a result of food riots. Unless a solution is found, the situation is likely to become worse as the century progresses, because it is projected that by 2050 there will be 9 billion mouths to feed, as opposed to today's 6.6 billion. Much of the remainder of this essay will be devoted to the nexus between carbon sequestration and food production, for it's here that we interact powerfully with the carbon cycle in ways that make it possible to achieve climate security and food security simultaneously. I will argue that it's possible to increase the yield of agricultural and pastoral land while at the same time sequestering carbon. In effect, we can create an ecological "magic pudding" which, while not infinitely elastic in its capacity to feed us, can stretch much farther than we commonly suppose.

An impressive amount of environmental experimentation is currently going on in agriculture, including a zero-based approach to crops and the establishment of permaculture. Most of these innovations produce incremental gains in productivity, as well as some carbon sequestration. One class of technology, however, promises a solution different in both quantity and quality from all the others. Known as pyrolysis, it generates energy, improves soil, and permanently withdraws carbon from the atmosphere, all at the same time. Pyrolysis is an everyday phenomenon that involves the heating of biological matter in the absence of oxygen. It occurs when the outer layer of the biomass oxidizes (burns), but the inside does not, as in frying, roasting, and toasting. For instance, because oxygen can't reach inside the steak you cook on the BBQ, it cooks by pyrolysis. Charcoal—composed chiefly of carbon—is a product of pyrolysis; making charcoal by such means is a practice that goes back thousands of years. It's also what kept cars on the road in places like Australia during the World War II: they ran on charcoal burners. Modern pyrolysis is simply a very sophisticated means of making charcoal, but in its end result it is far different from any pyrolysis known previously. Any biological material — crop waste, animal manure, forestry offcuts, and even human sewage — can be used as a feedstock.

Why is charcoal important? When we grow trees to offset the carbon emissions created by the burning of fossil fuel, we are really trading a very secure form of carbon sequestration (that oil or coal would have stayed buried for millions of years if we hadn't dug it up) for a less secure form of storage. After all, the carbon in a tree is "volatile" in the sense that if the tree rots or burns, the carbon will quickly be released back into the atmosphere; and climatic, political, and economic changes can all make the tree likely to burn. All living things are made of volatile carbon: you and I, for example, will rot away quickly when we die. But when we turn biomass into charcoal, we transform much of that carbon from something volatile to something inert. Plowed back into the soil, charcoal won't readily rot or burn, a fact that scientists exploit when they use ancient charcoal for carbon 14 dating. So when we turn biomass into charcoal, we're effectively

79

breaking the carbon cycle—by locking CO_2 that the plant took from the atmosphere as it grew into a solid material that cannot rot and cannot return to the atmosphere for hundreds, even thousands, of years. (This is also true when the biomass is an animal that ate the plant.)

Let's examine what happens to a crop of corn in the normal course of events. After the cobs are harvested, the crop waste (called corn stover) is usually left to rot in the field. Rotting doesn't take long: the bulk of the carbon in the crop waste returns to the atmosphere as CO_2 in a few months. If, however, the farmer harvested the entire corn plant and processed the nonedible portion through a pyrolysis machine, one-third of the carbon captured by the crop would be transformed into charcoal, and the rest would be incorporated into a synthetic gas or oil, which could in turn be used to generate electricity or transport fuel.

There are several kinds of pyrolysis machines. Some are classified as "slow": the biomasss spends a long time in the machines, which create synthetic gas. Others are "fast": they briefly subject biomass to temperatures of up to 930 degrees Fahrenheit, and they can produce a bio-oil or a gas. Pyrolysis machines also operate at various scales, from those capable of processing just 110 pounds per hour to those that process up to 4.4 tons per hour — and the machines can be made transportable.

Some of the synthetic gas generated by the process is required to run the machine; but if the crop waste is dry there is a lot left over, which can be used to generate heat or electricity. Where farms are connected to the electricity grid, and legislation permits the transfer of electricity from renewable energy producers into the grid, this represents a major financial benefit to the farmer. The real biological benefits, however, occur when the charcoal is plowed back into the field.

Charcoal is very porous — after all, it was once living cells — and its pores contain residual nutrients and minerals. If it is plowed back into a field, bacteria and soil fungi essential to healthy plant growth soon colonize its cellular spaces, and this colonization leads in turn to healthier soil flora and fauna. That effect, along with charcoal's capacity to lower soil acidity and retain moisture, generally results in a better crop the following year. Charcoal also holds moisture, and because of the soil's improved moisture retention plants have longer access to any fertilizers that are applied to the field. Also, the fertilizers tend not to run off into creeks and rivers, so water quality is improved. Of course, different soils react differently to applications of charcoal, but overwhelmingly the indications are that most soils benefit from it.

A study published in 2007 in *Nature* indicates the immense capacity of pyrolysis to change our world. By pyrolyzing current crop and forestry wastes in the United States and plowing the charcoal into farm soils, we could offset 10 percent of the nation's fossil fuel emissions. If this practice was vigorously pursued on a global scale, by 2030 humanity could be pulling an estimated nine gigatonnes of carbon per year out of the atmosphere using pyrolysis machines. As already noted, the standing stock of carbon pollution in the atmosphere is roughly 200 gigatonnes. Nine gigatonnes represents nearly 5 percent of this. If the carbon sequestered through tropical forestry is added to that, it would amount to a mighty contribution to combating climate change—all told, between 7.5 percent of the standing stock *per year*. And pyrolysis offers even greater greenhouse benefits, for it greatly reduces the production of agricultural nitrous oxide, which is 270 times as powerful a greenhouse gas as CO_2. More will be said about this gas soon.

Since pyrolysis has so many benefits, why is it not more widely utilized? Pyrolysis machines are expensive, and farms are mostly still family businesses. If farmers are ever to be able to afford these machines, they'll need to be paid approximately thirty-seven U.S. dollars per metric ton for the carbon they create, most likely through carbon trading schemes that recognize charcoal as a method of carbon sequestration.

Animal Solutions

The world's rangelands are too dry, or their soils too poor, to support agriculture, yet they far exceed in extent the area of arable land. Today, the more marginal rangelands are also under threat of severe degradation through desertification and soil erosion, as we see in the Sahel region in sub-Saharan Africa and in northern China. Such degradation results in the release of carbon from the soil. It is difficult to estimate the contribution such desertification makes to the atmospheric carbon flux but probably the effect is considerable. For this reason, even small increases in the rate of soil carbon capture on the rangelands can have a large positive impact. Unfortunately, however, soil carbon is an area in which there are few precise studies of what can be achieved by way of rapid carbon storage. Rattan Lal of Ohio State University, who is one of the world's most eminent soil scientists, estimates that by using existing technologies and practices to improve rangeland management, humanity could pull approximately one gigatonne of carbon from the atmosphere per year. Of course, the soil carbon thus sequestered is volatile (it is likely to be lost again if erosion recommences), but this is still a valuable contribution.

Worldwide, many graziers are radically rethinking the nature of their businesses in ways that promise to enhance soil carbon significantly. One of the most promising approaches, which is now being practiced on about 12 million acres globally, is known as holistic management. Holistic management was devised by Allan Savory, a white farmer who as a young man fought in the Rhodesian civil war against Robert Mugabe's forces. Spending long periods in the bush, Savory noticed that grazing animals bunched up tightly and moved rapidly from place to place as a defense against predators. As a result, grasses were eaten down severely over a brief period, but then were left for a long time to recover. The behavior of wild herbivores observed by Savory was not markedly different from traditional grazing practices: herders kept their stock in tight groups, and on the move. But during the nineteenth and twentieth centuries, when barbed wire came into use, pastoralists began enclosing pastures in paddocks and dividing the live-stock so as to control them more easily—for example, cattle were divided into breeding cows, steers, and bulls. This was disastrous for the pasture. Under such circum-stances, cattle selectively eat the sweetest feed, and this preference gives the less palatable plants an ad-vantage. Soon, farmers find that their pasture is in-vaded by shrubland or weedy species, which severely limit the productivity of their herds.

Over the past few decades some pioneering gra-ziers, following Allan Savory's methods, have been doing things very differently. They use electric fenc-ing to divide their land into small paddocks, and put all their livestock together in just one of these pad-docks. Within a day or two, the cattle have eaten every-thing they can reach, and the ground has been churned into a bare, dung-filled mass. The farmer then releases the cattle into the next small paddock, and so on until, at harvesttime, the cattle enter the pen next to the stockyards, whence they can be taken with minimal effort and fuss to market.

The result of this practice is that the weeds as well as the best pasture grass on the property are eaten, and as the plants regrow in the dung-enriched soils, those that put less energy into chemical defenses (which make them unpalatable) are able to colonize the greatest area. These species, lacking chemical defenses, are of course the sweetest feed for cattle, so in a single blow the battle against the weeds is won. Farmers following holistic management also find that they need to use less medicine, because the cattle that are not sent to market do not return to the same small paddock for a year or so, and this gap breaks the reproductive cycle of many parasites. Even more impressively, farmers are able to increase the number of stock they hold, often to an astonishing extent. I've visited farms practicing holistic management that have more than seven times the number of cattle they could feed using conventional methods. Just imagine what this means for farm economics: 700 healthy cattle where there were previously just 100. And the land on such farms is also much better off: trees thrive in the paddocks; and native species such as kangaroos, wallabies and birds are present in abundance, drawn to the green plants and fertile soil.

What is the potential carbon impact of such practices? Savory is considerably more optimistic than Lal about the potential of rangelands management to sequester carbon. He notes that about 4 billion hectares of the world's rangelands, including most of the world's dry rangelands, are threatened by degradation, and that merely preventing desertification and erosion would lead to a significant abatement of greenhouse-gas emissions. But where holistic management is practiced—at least on the better soils—increases of up to 3 percent in soil carbon are being achieved. This happens largely because the soil is protected from erosion and grass cover increases, allowing more root growth. Although the result in terms of gigatonnes remains in question, there's little doubt that such an approach could be a powerful way to cleanse our atmosphere of excess greenhouse gases.

There are other ways to sequester carbon in the world's rangelands, and one method, which is being utilized with increasing success in protected areas, is to alter the timing and scale of wildfires. If tropical rangelands are burned late in the dry season, the fires are so hot and widespread that they reduce soil carbon and bare soils over huge areas. The Australian Wildlife Conservancy, which controls about 6 million acres of preserved land in Australia, is pioneering a method of burning during the early dry season that preserves carbon in the soil and has substantial benefits for wildlife. This means that even in conservation areas such as national parks, significant soil storage of carbon can occur.

As beneficial as these technologies are, it would be simplistic to add the figures up and claim that the "historic debt" could be repaid using them. Even if all of the 200 gigatonnes were drawn down by 2100, the problem would not be solved, for some of the CO_2 that has been absorbed by the ocean would then be released because the partial pressure of CO_2 at the ocean's surface would change. A 200-gigatonne drawdown would, however, represent a stunning advance toward a stable climate.

It is argued by those who oppose eating meat that cattle produce methane, and that therefore a better strategy would be to destock the rangelands altogether. But is it really desirable to abandon use of the world's rangelands at a time of perilous food security? Furthermore, if the rangelands were to be destocked and left unmanaged, it is likely that fire would burn the vegetation, and as a result more carbon would enter the atmosphere and there would be huge increases in nitrous oxide.

I believe that in a world facing a food shortage and a climate crisis, livestock represent a potent weapon in the fight to stabilize our climate. Another consideration is an astonishing advance in livestock management that has recently been made in New Zealand. Cattle and sheep produce methane and nitrous oxide (N_2O), and their impact is so great that half of New Zealand's greenhouse-gas emissions come from livestock alone. Approximately one-third of these emissions are N_2O, generated because cattle and sheep eat plants that contain nitrogen in excess of their needs. The excess is excreted in their urine, and when it reaches the soil, bacteria transform the nitrogen to N_2O. The same thing happens when excess nitrogen fertilizers are applied, resulting in considerable waste because the nitrogen (as N_2O) moves into the atmosphere and is no longer available to plants.

There are several ways to deal with this problem: supplementing cattle feed with fodder less rich in nitrogen can make a difference, as can encouraging the cattle to spend time on "cattle pads" (areas of straw or bark where the urine can be treated). But by far the greatest benefit comes from treating pasture with nitrification inhibitors. Indeed, this method is so

extremely effective that applied in one sector alone—
dairying—it could reduce New Zealand's N2O emissions by 70 percent. And because it reduces the need
for costly fertilizers, it actually saves the farmer money.
New Zealand dairy farmers typically spend fifty-six
New Zealand dollars per acre applying nitrogen to
their pasture, but if they spent just forty-eight dollars
per acre on nitrification inhibitors, they would be
saving eight dollars, because they would no longer
need the fertilizer. Inhibitors are already in use in
pelletized fertilizer applications, so their use on farms
is not novel. Yet their potential impact is so great that
the $1 billion (approximately) owed by New Zealand
for excess emissions under the Kyoto Protocol could
be paid in toto were nitrification inhibitors used by
its dairy industry alone.

You might wonder why, if all this is true, every gra-
zier in the world is not applying holistic management
and using nitrification inhibitors. But farmers are
conservative: they generally follow what Dad did, and
will not quickly change. Some are on the land for the
lifestyle, and are seduced by the romance of the
roundup. Others think that holistic management is
too much hard work—yet in this they are wrong, for

over the course of a year the workload is roughly equal to that of the traditional model. However, it's more steady (it involves monitoring the pasture and opening the fence every few days for the cattle to move on), rather than requiring a huge effort at roundup time.

Farm-Based
Ecological Efficiency

In my grandparents' time, mixed farming was the predominant model. Families worked smallholdings to produce both food plants and animals. But the recent history of human food production has involved increasing specialization: over the course of the twentieth century farmers tended to increase the size of their holdings, and to specialize in growing a single crop over a large acreage using sophisticated farm machinery. This has been a bad thing in many ways, both for Earth's ecosystems and for our capacity to feed ourselves. That's because almost all of Earth's ecosystems depend upon the relationships between animals and plants in order to function. Indeed, plant-animal interactions are at the heart of Gaia's self-regulation. Plants capture the sun's energy, and animals, by feeding upon plants, create and swiftly recycle nutrients that plants need in order to grow.

A few pioneering farmers have realized this and have begun to integrate a wide variety of plants and animals into highly productive and sustainable enterprises. While researching his book on U.S. food production, *The Omnivore's Dilemma*, Michael Pollan visited Polyface Farm in the Shenandoah Valley in Virginia. The farm is owned by Joel Salatin, whose father purchased it in the 1960s when it was a dilapidated dairy farm full of erosion gullies. As a result of careful ecological sculpturing, that land now produces 360,000 eggs; 10,000 broiler hens; 800 stewing hens; 25,000 pounds of beef; 25,000 pounds of hogs; 1,000 turkeys; and 500 rabbits every year. All this is achieved without external inputs such as fertilizer, hormones, or antibiotics. Even more remarkably, Salatin has converted 400 of his 550 acres to forest. The birds and other creatures it harbors, and the wood chips it yields, play vital roles in maintaining the productivity of the 150 acres of worked land, and of course this forest also sequesters a significant amount of carbon.

The secret of Salatin's astounding success is a deep understanding of ecology. Everything on the farm works with everything else to promote fertility, and the creatures all seem to lead fulfilling lives. The laying hens, for example, roost in "eggmobiles" that are moved around the farm, allowing them to harvest the maggots out of cow pies and spread their own nitrogen-rich fertilizer around. The cows graze grass in a rotational pattern, sweetening the pasture, while the pigs root in the compost, helping to turn it over, aiding fertility.

At Polyface, Pollan slaughtered chickens, an experience that led him to investigate animal rights. He deplores the practices to which animals are subjected on factory farms and accuses such farms of cruelty. Yet after seeing how the animals live at Polyface and understanding how integral they are to the fertility of this marvelous farm, he decided that it would be better for the environment, human health, and for animals as a whole if most of us became eaters of sustainably produced meat. Many readers will doubtless ask whether vegetarianism is an even better option. Pollan reminds us that even vegan lifestyles involve cruelty to animals: think of the thousands of field mice shredded by harvesters, the woodchucks crushed in their burrows by tractors, and the songbirds poisoned by pesticides when farmers grow the wheat for our bread, he says. Pollan's message seems to be that to live we must kill, and the best we can do is kill humanely.

An important aspect of the success of Polyface Farm is Joel Salatin's belief that everything has its own scale. He doesn't want to have a bigger farm, or to produce more eggs (despite the fact that demand is high), because to do that would knock the entire enterprise out of balance. You need just so many chickens per cow on such a farm, and if the whole thing becomes too big, the farmer is unable to give the entire complex system the attention it requires.

One of the most heartening things about Salatin's farm is that all of the produce is eaten locally: this arrangement minimizes the use of fossil fuels in long-distance food transport. Most of the food we eat is likely to have traveled from half a world away. Local, sustainably produced food is likely to have less environmental impact, and to be fresher and of superior quality. "Oh, those beautiful eggs!" says one chef supplied by Polyface Farm. "The difference is night and day—the color and richness and fat content. There's just no comparison. I always have to adjust my recipes for those eggs—you never need as many as they call for."

These examples of changes in human food production provide a glimpse of what may be in store for us over the course of the twenty-first century. And such innovation is desperately needed. Fifty years ago, there were approximately 2.5 acres of arable land for every person on Earth, and the leaky, wasteful, polluting agricultural practices that had remained unchanged for thousands of years were adequate. But by the middle of this century, there will be just half an acre of arable land per head. Unless we find ways to use that land sustainably and creatively, humanity has no future.

It will not have escaped the attentive reader that many of the solutions outlined above relate to meat production. How can this be squared with the widespread view among environmentalists that eating meat is unsustainable and polluting? Conventional, high-intensity meat production, such as the production of cattle in feedlots, is indeed highly polluting and unsustainable. The fact that such cattle are fed grain, which is grown using fossil fuels, and that they transform just 10 percent of its food value into meat, demonstrates the dangers. But clearly not all meat is grown that way. Holistic management and mixed-farming practices can produce an abundance of meat and at the same time result in the sequestration of carbon from the atmosphere. Furthermore, for the world's dry rangelands there seem to be few, if any, cost-effective alternatives to holistic management.

Despite the advantages of sustainable meat production, I fear that the times are very much against the development of such solutions, for there is a growing feeling in western society that eating meat and owning livestock are morally wrong. This position, amounting to an ideology, is based partly on a belief that it's wrong to kill animals, partly on a belief that eating meat is unsustainable, and partly on health concerns, which, in my opinion, add up to little more than faddism. Indeed, those attracted to food fads today have a plethora of options—vegetarianism, veganism, and fructarianism being just three. What we really need in the twenty-first century, I believe, is a different approach to food. We should be eating what is good for the planet, as well as what is good for ourselves—a sustainabilitarian diet. Such a diet could, of course, also be vegetarian, vegan, or any kind of "arian" you wished, so long as the food eaten was sustainably produced.

Today, those who wish to adhere to a sustain-abilitarian diet need to research the origins of their food in great detail, or else produce it themselves. The situation would be much easier for the consumer if a better labeling system for food was enforced. Such a system should state clearly the distance food has traveled and the practices used to grow it. It's an indication of the dubious origins of some of our food that no such system is in place yet. Clearly, the purveyors of factory-produced chickens and pork, unsustainably produced crops, and fruit subjected to long transportation and storage don't want us to know what we're eating. In a democracy, citizens' action can achieve such commonsense regulation. After all, what's at stake is the future of our health and our planet. And putting us in touch with the origins of our food may have more pervasive effects. It could shed light on cruelty to animals as well as unsustainable production, and ultimately it might help liberate us from the great human feedlot which imprisons most of us, and which is a basic cause of our wasteful attitude toward water, energy, and food.

THE AGE OF SUSTAINABILITY?

I think that there is now a better than even risk that, despite our best efforts, in the coming two or three decades Earth's climate system will pass the point of no return. This is most emphatically not a counsel of despair; it is simply a statement of my assessment of probability.

It is also an overture to a plea for consideration—if the times ever require it—of a desperate measure: fighting greenhouse-gas pollution with another kind of pollution. Imagine a situation that might occur in coming years. The Arctic ice cap is gone, and the Greenland ice cap has suffered a partial collapse, raising sea levels by almost eight inches. Now another collapse is imminent—one that might submerge London and Shanghai. We need an immediate fix to cool the planet, one that we can implement in a few months rather than years.

The Nobel laureate Paul Crutzen has considered this problem and come up with a possible solution. It

is based on the observed impact of very large volcanic eruptions, which have produced an immediate
global cooling by ejecting sulfur and ash into the
stratosphere, producing periods that were known in
Europe as "years without summer." Rather than waiting for an eruption to cool Earth, Crutzen argues that
we could use the world's jet fleet to administer a
measured dose of sulfur to the stratosphere to cause
global dimming. Modern jets fly in the lower stratosphere, and sulfur released there would remain air-
borne for a considerable period. The dose, of course,
would have to be sufficiently large to offset the greenhouse gases, and sustained over a long enough time
to avert the melting.

The risks of this strategy are, arguably, high. Sulfur dioxide can destroy ozone, and our ozone layer
is already compromised. Furthermore, it could vis-
ibly alter the sky, affecting sunrises and sunsets and
possibly changing the sky's color. Were we to exer-
cise this option of last resort, we would be living in
what looked from space to be a duller world—a
world akin to what you see under the haze that domi-
nates Beijing and Shanghai. But if all else fails, who
are we to say no to a strategy that exchanges a little

of Earth's natural beauty for our survival? Why, you might ask, have I raised this possibility now? Simply because if we hope to see the step taken in the future, we need to discuss its merits and disadvantages now. The process of assessing such drastic interventions in Earth's climate system is currently under way in Australia's premier science organization—CSIRO. But the issue should be on our political and social radar screens as well.

Anyone reading this essay might be overwhelmed by the scale and the number of the challenges facing humanity and wonder whether it's too late to avoid catastrophe. Here I've focused on our most urgent crisis—the climate problem—because I believe that only by setting priorities and devising solutions to address multiple problems simultaneously can we secure our future. But given our overall lack of awareness of the climate crisis, the nature of our political response thus far, and the limitations of our economic system, can we possibly avoid disaster? I believe that this question does not have a yes-or-no answer. After all, even if some catastrophic consequences are inevitable, their scale and the speed at which they arrive are still at least partially within our power to influence. And of course the potential for the greatest catastrophe—a new dark age following the breakdown of our global civilization—lies entirely with us, for it will occur only if we fall to fighting among ourselves.

At the beginning of this essay I suggested that sustainability is essentially about extending the Eighth Commandment to forbid stealing from future generations. What kind of society is likely to value the lives of those yet to be born to such an extent that it will sacrifice a little present wealth in order to assist them? Clearly, there is a relationship between how we value ourselves and our fellow members of society, and how we value the generations to come. Societies which treat their members fairly in law, which seek to eliminate poverty and great inequalities of circumstances and wealth, and in which care—perhaps one could use the word love—for one another is manifest in day-to-day life, are surely best equipped to grant to future generations their just consideration, and so deal with the great challenges of this century.

During the twentieth century a very different social model flourished. Slogans such as "survival of the fittest" and "greed is good" expressed a belief in the virtues of a society where greater value was placed on individual enrichment than on the well-being of our fellows. At their most toxic, these ideas came together in "social Darwinism," a belief system that would have appalled the great architect of evolutionary theory. It is often distilled into a dog-eat-dog doctrine: "We'd better keep others down, keep growing, and remain strong, because if we don't, we'll be attacked and destroyed." To the adherents of social Darwinism, the world is a deeply hostile place, where a nation or individual must remain in control or face destruction. It's a paranoid, self-fulfilling philosophy, which perhaps more than anything else threatens to rob our children of a future. In a perverse way, perhaps the old guard—the Cheneys and Bushes of the world—were right to so resolutely oppose action on climate change, for there is no place for their ideologies, economies, or wars in the world we are about to enter. Like the generals of old, they may have preferred to go down defiantly in a world racked with conflict.

A sense of hopelessness is just as great a danger to our future as these bankrupt philosophies. Our world abounds in millennial cults for whom the "last days" are close at hand. Who, holding such a belief, would strive to save the world? Even world-weariness, a resignation to destruction, is profoundly inimical to sustainability, because its adherents believe that the fate of our planet is already sealed. If the British had thought that way in 1941, we might be living in a very different world.

We citizens of the developed countries bear a special responsibility in this world of imbalance among Gaia's organs, for we are the greatest gougers at the Earth—those who freed the carbon that now, like a malign genie, threatens the entire world. With us lies the bulk of the burden of ensuring that whatever we unearth does not, as it disperses into the waters and the heavens, destroy the balance upon which life depends.

If we are successful in finding a sustainable way of living in the twenty-first century, then perhaps the principles we develop will become the guiding principles of a truly sustainable global civilization. Whatever the case, increasing awareness of our unique position and role on planet Earth will necessarily drive political, economic, and social agendas long after our current preoccupations have faded.

I believe that each century has its own unique challenges, which if met, breathe life into the century that follows. In the nineteenth century it was social injustices that presented the greatest threat to humanity's future. At the dawn of that age, it was perfectly legal and acceptable almost everywhere to own another human being. Appallingly cruel child labor was entirely unexceptional, and only a few wealthy men had any say in who governed. The abolitionists, unionists, and suffragists, fought a century-long battle to end these injustices and bring dignity to an ever-widening proportion of humanity; and without their efforts in the face of daunting opposition, we might still be living in a patriarchal, strictly hierarchical, slave-owning world.

The twentieth century had other concerns. By then, we had become capable of manufacturing weapons that could destroy all human life at the push of a button. The best of us worked on building the structures—such as the UN, the EU, and anti-ballistic missile treaties—that would keep us at peace rather than at total war. This, and the great medical triumphs, such as the banishing of smallpox, will be remembered as the triumphs of that age.

This twenty-first century of ours will be faced with appalling social injustices, conflict, and pestilence. But these will not be its defining challenge. Instead our task is a far more difficult one: to bring sustainability to a species that has not known such a condition since it manufactured its first tool. This is a defining responsibility, for by our actions we shall determine whether Gaia will achieve intelligent control, or whether the blind watchmaker will be allowed to tinker on with his tools of variation and sterility, just as he has for the past 4 billion years. If we fail, all of our species' great triumphs, all of our efforts, will have been for naught. And perhaps the last 4 billion years will have been for naught as well.

NOTES

2 **all plausible projections indicate**: See *World Population Prospects: The 2006 Revision* (2007), United Nations, Department of Economic and Social Affairs, Population Division, Working Paper no. ESA/P/WP 202; and M. Wackernagel et al. (2006), *Living Planet Report*, WWF, Gland, Switzerland.

16 **the largest and most definitive study yet on this subject**: C. Rosenzweig et al. (2008), "Attributing Physical and Biological Impacts to Anthropogenic Climate Change," *Nature* 453 (15 May), pp. 353–357.

18 **Scientists postulate**: See "And Life Created Continents . . . ," *New Scientist* 2544 (24 March 2006), a review of an article by M. T. Rosing, D. K. Bird, N. H. Sleep, W. Glassey, and F. Albarede (2006), "The Rise of Continents —An Essay on the Geologic Consequences of Photosynthesis," *Palaeogeography Palaeoclimatology Palaeoecology* 232, pp. 218–233.

21 **the entire volume of the oceans is capable of sustaining life**: Tony Koslow (2007), *The Silent Deep: The Discovery, Ecology, and Conservation of the Deep Sea*, University of New South Wales Press, Sydney.

27 "It is almost as if we had lit a fire to keep warm": James Lovelock (2006), *The Revenge of Gaia*, Allen Lane/ Penguin Books, London; quote from p. 9.

35 "We . . . are no longer trying to protect the Arctic": "The Arctic: Is Dangerous Climate Change upon Us?" presentation by Neil Hamilton at the Australian Academy of Science's 2008 symposium, *Dangerous Climate Change: Is It Inevitable?* Canberra, 9 May.

38 "Look out on the surface of the great sea itself": Extract from Peter Ward (2007), *Under a Green Sky: Global Warming, the Mass Extinctions of the Past, and What They Can Tell Us about Our Future*, Smithsonian Books/Collins, New York.

39 They looked back over: James Hansen et al. (2008), "Target Atmospheric CO2: Where Should Humanity Aim?" See <http://arxiv.org/abs/0804.1126> and <http:// arxiv.org/abs/0804.1135>.

45 a study examining the IPCC projections: R. Pielke, T. Wigley, and C. Green (2008), "Dangerous Assumptions," *Nature* 452 (3 April), pp. 531–532.

47 A commentator on this groundbreaking research: Editorial (2008), *Nature* 452 (3 April), pp. 503–504.

66 Although these rain forests cover: See P. W. Richards (1996), *The Tropical Rain Forest*, Cambridge University Press, Cambridge.

68   When Alfred Russel Wallace visited the island of Singapore in 1862: A. R. Wallace (1869), *The Malay Archipelago*, Macmillan, London.

82   A study published in 2007 in *Nature*: See Johannes Lehmann (2007), "A Handful of Carbon," *Nature* 447 (10 May), pp. 143–144.

88   But where holistic management is practiced: Information supplied by A. Savory, personal communication, May 2008.

91   New Zealand dairy farmers: Simon Terry (2007), "A Convenient Untruth: Towards a Lighter Agricultural Footprint," report published by the Sustainability Council of New Zealand, 25 June 2007.

96   Oh, those beautiful eggs!: Michael Pollan (2006), *The Omnivore's Dilemma: A Natural History of Four Meals*, Penguin, New York; quote from p. 252.

RESPONSES

BILL MCKIBBEN

Tim Flannery's book, and his title, couldn't be more correct. Since I had the dubious pleasure of writing the first book on global warming twenty years ago this fall, I've been able to watch the whole process unfold: from hypothesis to scientific consensus to a kind of panic in the last two years as the pace of change speeds up and it becomes apparent that we're in for much bigger, faster change than even pessimists bargained for. And in case you think Flannery is being unduly alarmist, you should know that he barely even starts down the list of the massive planetary changes unleashed by our infusion of carbon into the atmosphere. Had he wished, he could have taken you on a tour of the world's high-altitude glaciers, which supply much of the world's drinking and irrigation water and are now melting away to nothing. He could have described the recent droughts now turning large parts of the planet (including his native Australia) to cracked mud. He could have described the most recent studies showing that food supplies will be difficult or impossible to maintain as the century wears on. He could have—but why bother? If a purple ocean under a green sky doesn't make the case sufficiently to move you to action, it's not clear what will.

I wish to suggest—here in the late summer and early fall of 2009—something that you can do to help slow this crisis down. I am here to supply you with an action that you can take, one that will be I think more than pyrrhic, even if less than totally successful. But first, a little analysis of the ground we're fighting on. Flannery describes a set of possible technologies, such as pyrolysis and rechargeable cars, that he thinks will be useful in dramatically reducing carbon emissions. The question he doesn't completely address is: how can we make them happen quickly enough to matter? That is, everyone knows we'll be driving nifty eco-cars (if we're driving anything at all) in 100 years. What we need to know is: how can we make it happen in ten?

There's only one lever even possibly big enough to make our system move as fast as it needs to, and that's the force of markets. It wasn't until gas hit four dol-

lars a gallon last summer that Americans suddenly
began to reconsider the SUV as an object of their af-
fection. If we could make the price of fossil energy
consistently high, then we might be able to get everyone (even those who haven't read this fine book) to
change his or her habits: to drive smaller cars or take
the bus or bike or walk or just stay at home; to build a
smaller house, or move in with your mother, or rent
an apartment near your job; to eat locally, and lower
on the food chain, and to grow your own food; to stop
trying to meet nonmaterial needs (love, respect, affec-
tion) by buying stuff at the store. All of us in a consumer society have fallen into these kinds of habits
because they are affordable, and they are affordable
because fossil fuel doesn't bear the cost of the damage it does to the environment. Until that changes,
nothing really will change. You can't make the math
work one lightbulb at a time.

But governments have to take the step to make
that happen — have to pass the caps on carbon that
will make coal and gas and oil carry a price. And governments are reluctant to do that. They're reluctant
because vested interests carry great sway in their de-
liberations (Exxon Mobil made more money last year

than any other company in the history of money). And they're reluctant because they fear being punished by voters if the price at the pump rises. So we sit, immobilized. The world's leaders will go to Copenhagen in December 2009, as Flannery points out. If they don't take strong action, then our last plausible bite at this apple will have passed—our last real chance to rewrite the economics of carbon in time to prevent the worst catastrophes.

So—we need to move governments. And the way to move governments is to build a real citizens' movement that demands change. That is why you need to put this book down in a few minutes, go to your computer, and visit 350.org. Beginning the day that Jim Hansen published the paper described by Flannery in this book—the one that set the red line for the atmosphere at 350 parts per million—a few of us launched the 350.org campaign. Its goal is simple: to take the most important number on Earth, and make it the most well-known number on Earth. This effort—led mostly by young people—spent a year building real support among the cognoscenti: everyone from Tim Flannery to the Indian activist Vandana Shiva to the great Canadian environmentalist David Suzuki agreed that this number would be the rallying cry, the first time that the world has tried to rally not around a slogan but around a scientific fact. It's arcane — on the other hand, Arabic numerals translate across languages, which is a great help on a globe where people insist on speaking in their own tongues.

For the next few months, we're pushing as hard as we possibly can to make that number famous. We've set a date: October 24. It will be an international day of action designed simply to spread that number. There will be thousands of events around the globe — climbers with banners high in the Himalayas, 350 scuba divers on the Great Barrier Reef, teams hanging banners from the stone guys on Easter Island. Churches ringing their bells 350 times, people planting 350 trees, groups of 350 bicycles circling the center of town. Pyramids of 350 pumpkins outside your farmers' market. Anything that will allow you to educate your neighbors, and get on the front page of your local paper. You can do this. You do not need to be a professional organizer, or even an amateur — you need to have an e-mail account with the names of some friends in your address book. You send them a note, they send the note on, and pretty soon you've organized yourself an event. We have materials at 350.org to help you out, and all kinds of examples of the things people are doing all over the world.

We also need you to forward news of this plan around the globe. The one wild card we have in this fight is the availability of easy and quick communication—without the Internet we wouldn't have been able to think of this plan. This is what the Internet was invented for: not playing poker in your underwear all night, but spreading one particular piece of information to every corner of the globe, in time for it to matter. You, with your laptop, are more connected to the rest of the world than the most connected person on earth twenty years ago. If you want to do something about what Tim Flannery has described, then sit down for two hours tonight and write to people you know around the world, forwarding on our call for actions on October 24.

I can't promise this will work. As Tim has said, the momentum behind these physical systems is large enough that it's possible *nothing* will work. And the vested interests that want to delay action are very, very strong. But doing nothing, or waiting for Obama to do it by himself, or expecting some miracle technology to appear—those don't meet Flannery's definition of "mature." We may be, as he says, the agents of our own protection. But if that's the case, we need to act. An antibody that just sits there is a waste. An antibody that goes to work against the trouble is what we need.

BILL MCKIBBEN is a scholar in residence at Middlebury College and the coordinator of 350.org.

RICHARD BRANSON

With scientific gravitas, complemented by the skillful use of layman's language, Tim Flannery paints a serious picture of the planet's future, even if, as he says, he overwhelms us with "the scale and the number of challenges facing humanity."

Let me start with a huge dose of optimism. I believe that we will rise to the challenges Tim poses. I believe it is possible that one day we will enjoy modern, fun-filled lives using only one planet's worth of natural resources. We will emit minute amounts of carbon; there will be radically less evidence of poverty; and most people, most of the time, will enjoy healthy, satisfying lives. Sustainability is possible. If I have been successful, it's because I believe the impossible is possible and I have, in my business life, made it so. OK, building a business is not saving the planet, but we all need that same "Let's do it" attitude if we are going to see this challenge through.

As a businessman responsible for nurturing companies, careers, and customers, as well as meeting environmental and social responsibilities, I find the major challenges, particularly climate change and food production, almost too vast to contemplate. At the moment, the tone of voice around sustainability implies sacrifice and giving stuff up. Unsurprisingly, consumers reject this because it seems to present fewer opportunities for a satisfying life. We know that the opposite needs to be true. To this end, the world's experts could join us in a bigger debate about lifestyle choices and lifestyle possibilities.

None of us can afford to be reluctant to comprehend the scale of change required. We need to get ready for drastic as well as piecemeal action. I don't deny that we have all been guilty in the past of ducking the issue. Governments and politicians, by their nature, think in the short term, to the next election in democratic countries at least. Businesses have to meet the demands of their shareholders, who want short-term as well as long-term profit. Consumers— the general public—fret about the lives of their children or their grandchildren, but as individuals they feel powerless to do anything and question the difference it would make if they did.

I believe there is still a gap between the way business leaders think and the way environmental experts, such as Tim, think. Businesspeople consider the laws of economics, while Tim considers the laws of nature. These two sets of laws are not natural bedfellows. The fundamental challenge facing us all is to make the necessary and important rules by which we run our economy complement the laws of nature. The laws of economics were created in modern history to serve mankind, whereas the laws of nature go back billions of years and serve the entire ecosystem on which we rely. Therefore, while the laws of economics matter, they cannot overrule the laws of nature, and perhaps that is the humble pie Tim is inviting us to eat.

I am, however, heartened by a growing realization that businesses, governments, and citizens can form a powerful triumvirate to act in concert. While there are squabbles and disagreements, there is also a movement supporting the best academic and scientific brains, as well as admired statesmen, in the belief that "something has to be done." Initiatives such as the

Elders and the Environmental War Room are indications of this, each aiming to find solutions to global environmental and social challenges. The purpose of the Environmental War Room is to evaluate major solutions to climate change and to create incentives to enable their rapid and scaled deployment.

Tim Flannery deals with the macro issues facing society, and my businesses can and will make an important contribution to these. We will strive to understand where our products help provide short- and long-term contributions while reducing the negative contributions. It is complex. Because of the carbon emissions that result, it is easy to be harsh on the family flying to the Caribbean, but what about the benefit to the local communities and the benefit of quality family time? And what about the members of our health clubs—is their desire to keep fit a positive contribution in its own right? How can our mobile communications help rural and inaccessible communities? Patently, Virgin Group companies can make a positive or negative contribution toward making sustainable lifestyles easier, and each of them is being asked to identify what those contributions may be.

The questions we are asking our companies go beyond the usual corporate social responsibility puff and KPIs that some big businesses are expected to measure. Yes, we do measure our carbon footprint; yes, we do recycle and reduce waste; yes, we do invest in the latest, most fuel-efficient planes. But we also ask broader questions such as: What does an economy that uses only the resources of a single planet look like? How do we decouple economic growth from the use of natural resources? How do we contribute to lives powered by clean and renewable energy?

We are a business, so commercial success is foremost in our mind. But we also ask ourselves how we can ensure that the basics of the free-market economy will still operate, albeit with rules that are better aligned with the laws of nature.

Tim Flannery points out the power of tropical rain forests and their need for protection. The tropical rain forests are home to an estimated two-thirds of all living species; to hundreds of millions of people; and, as he emphasizes, to some of the world's most unsustainable agricultural practices. Like Tim Flannery, we believe that perhaps the biggest single opportunity that links the need to address poverty in developing countries and the need to reduce the rate of climate change is reversing the rapid and unsustainable rate of deforestation.

To do this, we need to ensure that rain forests are worth more alive than dead. So another question we ask ourselves is this: what influence can we have on developing creative ways of giving financial value to eco-services provided by rain forests and oceans to ensure that the economy will work within the finite limits of nature? At the moment, a rain forest generates more income when it has been converted into garden benches and oil palm. How do we create more income by leaving the original forest standing?

One of the Virgin companies—Virgin in the United Kingdom, together with Virgin Unite—has just begun working with the Climate Tree, an initiative of the Tropical Forest Trust, helping to finance a project in the Congo to find entrepreneurial ways of helping local people create value from their forests without causing damage.

Some readers might be bristling, annoyed by my focus on rain forests when airlines are meant to be one of the most evil perpetrators of climate change, but as a journalist from the UK *Independent* wrote last year, commenting on the UK Stern Report:

It is unwise for politicians to arm-wrestle over rising aircraft emissions when just the next five years of carbon emissions from burning rain forests will be greater than all the emissions from air travel since the Wright brothers to at least 2025.

It has also easily overtaken aviation as a source of greenhouse-gas emissions (500 million servers and growing). However, aviation is still seen as high profile and is coming under emissions-trading schemes. There are proposals in both Europe and Australia. Naturally I would prefer a single global scheme, but either way these schemes will generate billions of dollars. I believe that some of this money should be channeled into projects to protect rain forests.

It is not just about carbon and rain forests. Well-being is an important element of a sustainable lifestyle. A generally wealthier population in the western world has not always led to increased happiness and well-being. Instead there has been an increase in obesity and stress levels, as well as in diseases such as diabetes. So we are debating how our businesses can ensure that self-esteem and pleasure are based more on experience and the realization of one's potential than on the ownership of more and more stuff. We also ask how we can help people extend their personal well-being into community well-being. In many cases, the solution is simple: recycle more, keep fit, and buy greener products.

This does not discount the need to find large-scale technical solutions, and initiatives such as the Earth Challenge (a prize of $25 million to encourage a viable technology that will remove at least 1 billion metric tons of atmospheric CO_2 equivalent per year) will make hugely important contributions. These solutions, along with the development of nonfossil-generated energy and the major challenge of containing population growth, are all part of the bigger picture of true sustainability.

While I hope these questions we are asking show some robust intellectual thinking, I am also aware that more action on the ground is required. We will continue to work with partners more expert than us to ensure that we are tracking in the right direction; we will ensure that new investments contribute to, rather than work against, achieving sustainable lifestyles; we will resource our own experts and reinforce the knowledge of these issues among our senior managers; and we will encourage our businesses to be leaders in their sectors.

While the future challenges are massive, and at times the outlook seems bleak, I persist in seeing the glass as half full. To sit on the sidelines is to place our way of life at risk and possibly see millions of people die of starvation or suffer from extreme weather conditions. Such a prospect is what provides the impetus to act and to act now. No single group can solve the problem, and that is why we need to work together, whether as individuals, businesses, governments, or NGOs, to reach creative, pragmatic, yet bold decisions that will create tipping points for the challenges we face.

Some might think I am too optimistic. However, I would rather be optimistic and proved wrong than pessimistic and proved right. That's entrepreneurialism for you, and I know a little about that. Just imagine a world where the best scientists collaborate with the best entrepreneurs — perhaps then my optimistic vision will become reality.

RICHARD BRANSON is founder and head of the Virgin group of companies.

PETER SINGER

*No More Excuses**

Now or Never puts very well the urgency of the need to reduce greenhouse gas emissions. This book will, I hope, add to the already major impact that Tim Flannery has had in raising public and political awareness about one of the greatest moral challenges any generation has ever faced, and in spurring us to take the action that is so urgently needed. My view of what that action is, however, is a little different from Flannery's.

In making his case that we are indeed at a "now or never" moment, Flannery refers to the distinction

*This essay draws on a submission to the Australian government's Garnaut Climate Change Review, made in April 2008 by Geoff Russell, Barry Brook, and myself. Geoff Russell drew my attention to the significance of the time-frame against which we evaluate the role of methane in contributing to climate change.

drawn by James Hansen and his colleagues between climatic "tipping points" and "the point of no return." He then describes our present situation as one in which we are suspended between a tipping point — which means that the concentration of greenhouse gases in the atmosphere has already reached a level sufficient to cause catastrophic climate change — and a point of no return, at which the process leading to catastrophe will become irreversible. That image vividly portrays the momentous nature of the next few years in the history of humanity, and of our planet. As Flannery writes: "only the most strenuous efforts on our part are capable of returning us to safe ground. . . . There is not a second to waste."

I agree entirely. But this makes Flannery's failure to face up to the implications of eating beef all the more dismaying. For of all the ways in which people in affluent nations could rapidly reduce their contribution to climate change, ceasing to raise ruminant animals — essentially, cattle and sheep — is the one we could most easily achieve within the next decade.

Among those who have followed the debate about climate change, most now understand that rumination — which is involved in the digestive process of animals like cattle and sheep—produces methane, and that methane is a potent greenhouse gas. But few understand just how significant a role reducing the number of ruminant animals could play in helping us to avoid reaching the point of no return. This is largely because discussions about which human activities contribute most to climate change are usually framed in terms of the impact those activities will have over the next century. Taking that perspective, a ton of methane is generally regarded as twenty-five times more potent, in causing global warming, than a ton of carbon dioxide. That makes methane highly potent, but relative to carbon dioxide, this level of potency is heavily outweighed by the very much smaller quantities of methane produced by ruminants, compared with the quantities of carbon dioxide produced by, say, coalburning power stations. Hence methane emissions from ruminants are widely seen as being of much less concern than burning coal to generate electricity.

The reason why, over the next century, methane will be only twenty-five times as potent as carbon dioxide in causing global warming is that it breaks down much more quickly. Unless we find new ways of taking carbon dioxide out of the atmosphere, about a quarter of every ton we emit now will still be up there warming the planet in 500 years. But with methane, two-thirds of it will be gone in ten years; and by the end of twenty years, 90 percent of it will have broken down.

Suppose that instead of taking 100 years as our time frame, we asked which emissions will contribute to climate change over the next twenty years. Then the difference in breakdown becomes less significant, and a ton of methane is not twenty-five but seventy-two times more potent than a ton of carbon dioxide in warming our planet. That dramatically changes the situation in terms of which gases should be the target of our drive to reduce emissions.

Which time frame should we use, 100 years or twenty? Flannery has given us compelling reasons to choose the shorter period. If we have passed the tipping point and are approaching the point at which catastrophe becomes inevitable, there is little point in focusing on what impact the gases we are emitting now will have in 2100. As Flannery himself says, "There is not a second to waste." Twenty years is an amply long enough time frame because if we don't do something drastic by then, there will be no return.

Using the factor of seventy-two to convert methane to its carbon dioxide equivalent dramatically changes the balance between ruminant animals and coal-fired power stations. It implies that for some countries, cattle and sheep are the most important source of global warming. Australia's livestock, for example, produce 3.1 megatonnes (3.1 million metric tons) of methane. When we multiply 3.1 megatonnes by 72, we get 223 megatonnes of carbon dioxide equivalent—significantly more than the 180 megatonnes of carbon dioxide produced by Australia's coal-fired power stations.* Many other countries have very significant methane emissions from livestock, including Brazil, India, and the United States.

The importance of eating less meat, if we are to slow climate change, has been widely understood at least since 2006, when the United Nations Food and Agriculture Organization produced its report *Livestock's Long Shadow*, which said that livestock was responsible for more emissions than transportation. In 2008 Rajendra Pachauri, the chair of the IPCC, made an

*Australian Greenhouse Office, *National Greenhouse Gas Inventory 2005*, 2007.

explicit call to individuals, saying, "Please eat less meat—meat is a very carbon intensive commodity. . . . This is something that the IPCC was afraid to say earlier, but now we have said it."*

It is surprising, therefore, how little attention Flannery pays to what we eat, and how weak what he does say on that topic is. There is really only one paragraph in *Now or Never* that directly addresses the problem of methane from livestock:

> It is argued by those who oppose eating meat that cattle produce methane, and that therefore a better strategy would be to destock the rangelands altogether. But is it really desirable to abandon use of the world's rangelands at a time of perilous food security? Furthermore, if the rangelands were to be destocked and left unmanaged, it is likely that fire would burn the vegetation, and as a result more carbon would enter the atmosphere and there would be huge increases in nitrous oxide.

*www.abc.net.au/news/stories/2008/01/16/2139349.htm?section =world

Two points in this paragraph need to be addressed. First, what will happen to carbon emissions if rangelands are destocked? They may be more likely to burn, but after burning they will rapidly regrow, taking up carbon again. It isn't clear why this cycle will significantly increase greenhouse gas emissions for more than a few months, or a year or two at most. More important, in many areas forests have been cleared, and sometimes still are being cleared, to create pasture for cattle. Reducing the demand for beef will stop forest clearance and allow large areas of cleared land to return to forest, thus storing more carbon.

Second, to speak of "perilous food security" in the context of a *defense* of eating ruminant animals fails to take into account the fact that more than 750 million tons of grain is fed to animals each year, and a large proportion of that goes to cattle in feedlots.* (These cattle typically spend their early months on the rangelands, and their last months eating grain in feed-

*Food and Agriculture Organization (2008), *Crop Prospects and Food Situation*, No. 2 (April). Available at www.fao.org/docrep/010/ai465e/ai465e04.htm.

lots.) To get this into perspective, it amounts to more than half a ton of grain for each of the 1.4 billion people living below the World Bank's extreme poverty line — that's about three pounds per day, or more than twice as many calories as the average person needs. In addition, 80 percent of the world's soy crop is fed to animals. If we stopped eating meat, our food security situation would be far less perilous.

I am not suggesting that traditional herding people who have no real alternatives to eating ruminant animals should abandon their way of living. But the number of animals they have is tiny compared with the vast hordes of cattle, and to a lesser degree sheep, raised in the United States, Canada, Australia, and New Zealand. Eliminating these animals could be a major step toward slowing climate change. Moreover, it is something that is technically simple. Unlike phasing out coal-fired power stations, it does not require replacement by either a technology that already exists but is dangerous — nuclear power — or a technology that still needs to be invented, like solar electricity generation efficient enough to replace coal. We can cease to eat ruminant animals right now, and it will not bring our way of life to a halt. In fact we'll be healthier for it.

Flannery follows Michael Pollan in praising Joel Salatin's Polyface Farm, which has become something of an icon for the local, ecological farming movement. While I much prefer small, pasture-based farms to the dominant factory farms that lock animals indoors for their entire lives, my abhorrence of the latter is based largely on the wretched lives that billions of animals are forced to endure in the confinement and overcrowding of factory farms. If we are focusing on climate change, it is hard to see why factory-farmed chicken is worse than—or even as bad as—pasture-raised beef. Yes, like any form of factory farming, chicken production wastes food, because the food value of the chicken meat we eat is much less than that of the grain that the birds eat. It is also true that to produce the grain to feed to the chickens requires fossil fuel, so we are producing more greenhouse gases than we need to get the calories and protein we require. But given that methane is, over the crucial time period, seventy-two times as potent as carbon dioxide, beef is much worse for climate change than factory-farmed chicken.

Polyface farms produces, Flannery tells us, 25,000 pounds of beef per year. According to Ulf Sonesson, of the Swedish Institute for Food and Biotechnology, each kilogram of beef served is responsible for nineteen kilograms of carbon dioxide emissions, whereas each kilogram of potatoes served is responsible for only 280 grams — which makes beef sixty-seven times as carbon-intensive as potatoes.* It also means that Polyface Farms is responsible, through its beef production alone, for nearly 500,000 pounds of carbon dioxide emissions. In fact, that's probably an understatement, for two reasons. First, despite all the current enthusiasm about eating locally, when suburbanites think that a great way to spend their weekends is to drive out to the country and pick up their individual packages of ecologically produced food, many of the benefits of "locally" are blown away by the exhausts of their SUVs. That's what people who buy from Polyface Farm do. Look at the Polyface Farm website and you'll see the driving instructions — no public transportation is mentioned.† Sadly for those of us who would like to build connections between farmers and

*Janet Raloff, "AAAS: Carbon-Friendly Dining . . . Meats," http://www.sciencenews.org/view/generic/id/40934/title/ AAAS_Climate-friendly_dining_%E2%80%A6_meats.
†http://www.polyfacefarms.com/location.aspx

consumers, putting large quantities of food in a big truck and driving it to a supermarket near where people live is usually more fuel efficient.

The more serious problem for farms like Polyface is that from a greenhouse perspective, grass-fed beef is actually worse than grain-fed beef. Yes, you read that correctly. According to a study by Nathan Pelletier, of Dalhousie University in Canada, the greenhouse gas intensity of beef is roughly 50 percent higher when the animals are raised on grass than when they are finished on grain. This is largely because they eat a lot more fiber, and so their digestive system has to work much harder to digest it, producing much more methane as they do it.* Moreover, Pelletier's calculations were based on the usual 100-year time horizon. He has confirmed that using a twenty-year period instead would further increase the relative global warming potential of grass-fed beef over grain-fed beef.†

At a policy level, advocating a cut in cattle and sheep numbers ought to be a top priority, along with

*Raloff, "AAAS: Carbon-Friendly Dining . . . Meats."
†Nathan Pelletier, personal communication, 13 May 2009.

shutting down coal-fired power stations. Since politicians do not seem ready to take the necessary steps, every responsible environmentalist should lead by example. There are no excuses left for eating beef.

PETER SINGER is Ira W. DeCamp Professor of Bioethics in the University Center for Human Values at Princeton University and Laureate Professor at the University of Melbourne. His books include *Animal Liberation, Practical Ethics, Rethinking Life and Death, One World,* and, most recently, *The Life You Can Save.*

FRED KRUPP AND PETER GOLDMARK

Now We Are Called

When Tim Flannery says *Now or Never*, he means it.
And if we're smart, we will listen to him.

This essay on the immediacy and urgency of the need to address global warming specifically and our accelerating environmental imbalance generally comes from a scientist and writer who himself has only recently made the journey of inquiry and research that led him to these conclusions. In *The Weather Makers*, published in 2005, he describes the odyssey that led him to the conclusion that global warming was the paramount challenge of our time.

Now or Never is cast in accessible, commonsense terms, although it is informed by the training and discipline of a distinguished scientist. The message is not complicated: time is running out, Flannery tells us, and there are things we can do to reduce the chances of disaster. But the hour is late, the destructive

143

momentum of civilization-threatening emissions is enormous, and it is indeed "now or never."

Flannery's voice is clear and his argument compelling. But in terms of optimism his cup does not run over. He admits near the end of this essay that his personal feeling is that we will not rise to the challenge in time to avert the worst consequences of global warming. Part of the power of this compact essay lies precisely in the fact that he shows us that we can avoid catastrophe, and what it would take for us to do this. We are more optimistic than Flannery: we believe that when we invert the economic drivers, reverse the incentives, and harness private greed toward the right public objectives, we can summon tremendous entrepreneurial energy to drive emissions down or even soak them up; and that the history of entrepreneurism and technological change teaches us that this can be done more quickly, more easily, and at less cost than most defenders of the status quo would ever dream.

*

Flannery is a paleontologist and a mammologist, and some of the most fascinating sections of this essay deal with what we can do in the production of food to reduce carbon emissions and restore balance with our environment. Grazing rotation, use of nitrification inhibitors in dairying, and other approaches are explored. The suggestion that we can have food that is produced in environmentally sound ways, reduce cruelty to animals, and significantly curb carbon emissions at the same time is tantalizing and warrants immediate attention and rigorous exploration.

In one critical chapter, Flannery addresses what he calls "The Coal Conundrum." Coal is indeed the central knot in the tangle of problems and challenges that we face as we seek to curb carbon emissions before they unleash catastrophic and irreversible changes in our planet's ecosystems, food systems, climate patterns, and oceans. Because coal is the cheapest and most available fuel (as well as the dirtiest and most dangerous), both developed and developing countries have built much of their energy infrastructure on it. Flannery properly focuses on the need to get carbon capture and sequestration (CCS) systems up and running commercially at scale. Two of his conclusions are directionally unassailable: we will have to retrofit existing coal plants—there are too many of them supplying too much electricity, and countries struggling to develop are not about to knock them down; and this in turn means that the developed countries are going to have to pay for a significant share of this retrofit investment. Fortunately there has been significant progress on the technical front of this challenge — a fact Flannery does not cite in his bleakly despairing portrait of the leadership of the coal industry around the world.

Unfortunately, Flannery does endorse the ineffective "clean development mechanism," which so far has produced negligible net atmospheric benefits. Worse still, for the large emerging economies it acts as a *dis*incentive to join a rigorous global system of carbon limits. The coal conundrum is front and center, and we need a win-win solution that works for the developed world, the large emerging economies, the poorer developing countries, and the planet as a whole. To cut through this knot we will need policies, investments, and understandings across national borders that go beyond Flannery's essay.

Flannery is eloquent on the threat to our oceans. He tells us that the sea "has died" several times in the Earth's history, and that it can die again if we do not meet the challenge of climate change. His reconstruction of the "death of the oceans" that occurred in the past and how it can recur now as a consequence of global warming is vivid and terrifying.

The moral authority of Flannery's argument rests on his clear sense of human responsibility. He views the human adventure and the 4.5 billion-year history of our planet as tightly linked. We are the only creature whose activities and understanding have reached the point where the course of the former could influence the fate of the latter. And he places the primary burden on those of us who live, spend, consume, and pollute in the "developed" countries:

> We citizens of the developed countries bear a special responsibility in this world of imbalance among Gaia's organs, for we are the greatest gougers at the Earth – those who freed the carbon that now, like a malign genie, threatens the entire world. With us lies the bulk of the burden of ensuring that whatever we unearth does not, as it disperses into the water and the heavens, destroy the balance upon which life depends.

*

147

We write this essay as the U.S. Congress debates one of the most critical questions in its 222-year history: should the United States enact a strong cap-and-trade system to reduce carbon emissions? This will be a high peak for the United States to reach. From 1992, when the Rio Convention was signed, until 2008, the United States did virtually nothing at the national level to curb global warming. The history of assaults on unclimbed peaks suggests that there is usually a significant gap, a recuperation period, between a failed assault and its successor. That makes this current effort all the more pivotal. And it makes *Now or Never* all the more relevant and compelling.

FRED KRUPP is President of Environmental Defense Fund. PETER GOLDMARK is Director of the Climate and Air Program at Environmental Defense Fund.

GWYNNE DYER

Tim Flannery's greatest virtue is the clarity of his arguments, but he was insufficiently explicit and radical in his conclusions—quite possibly for tactical reasons, because they are implicit in what he does choose to say.

He rightly says that within the lifetimes of many readers, "Gaia will pass from an unconscious to a conscious means of control [of the climate]." He is talking about us, of course, and at the end of his essay he does refer to one deliberate human intervention in the climate that has gotten considerable publicity in the past two years: Paul Crutzen's proposal to inject sulfur into the stratosphere as an emergency preventive measure if global warming is getting out of control. But he does not admit (though I suspect he really knows) that direct human manipulation of the fundamental elements in the climatic equation—the amount of greenhouse gases in the atmosphere and the amount of sunlight reaching the

Earth's surface—is the way that things are going to be done from now on.

It is almost never acknowledged, in debates about how we prevent unfavorable climate change, that we are actually seeking to preserve one particular climatic state, desirable to human beings, out of a number of alternative possible climates that have prevailed in the past and may recur in the future—certainly will recur, in the case of another period of major glaciation, unless we eventually use our newly acquired ability to manipulate the climate to prevent it. Does anybody imagine that a successor civilization several thousand years hence, the beneficiary of our successful attempt in this era to avoid a global warming catastrophe, would not use the climate-control techniques we are developing right now to avoid a global cooling catastrophe as changes in the Earth's orbital pattern bring the current interglacial period to its natural end?

I may seem to be getting ahead of myself here, since it is far from certain that we will be successful in the present era in avoiding what would be, for human civilization, a catastrophic amount of global warming. Jim Lovelock is quite right to fear that a failure of political will could lead to a tenfold reduction in the human population by the end of this century. But we should be clear about the nature of our task: our agricultural and industrial practices, magnified by our huge rise in numbers, are driving the global climate in a direction that will hurt us very badly, and so our task is to change those practices in ways that drive the climate back into our preferred equilibrium. We are already manipulating the climate by our activities; success will be manipulating it in more intelligent ways in order to serve our ends.

Whether you want to dress that up as human beings becoming the consciousness of Gaia, or just see us as the same old self-serving species we always were, we are taking control of the planet's climate. This billions-strong human civilization will live or die by its success in understanding the global carbon cycle and modifying it as necessary to preserve our preferred climate. That is really what Flannery is talking about in his discussion of restoration of the tropical forests, the use of "bio-char" in agriculture, and the holistic management of rangelands: ways of bringing the atmospheric concentration of carbon dioxide back down below the ultimately disastrous level that it has *already* reached.

The consensus in climate-science circles is that we must never exceed a ceiling of three to four degrees Fahrenheit hotter, because somewhere between three and six degrees hotter we will trigger natural feedbacks, most notably methane releases from melting permafrost and a collapse in the carbon-dioxide absorption by the oceans, which would unleash runaway warming and remove the situation from human ability to control. Three or four degrees hotter is generally equated to an atmospheric concentration of 450 parts per million of carbon dioxide—but Jim Hansen's most recent estimate of the acceptable *long-term* concentration of carbon dioxide in the atmosphere, if we do not want all the ice on the planet to melt, is 350 parts per million. That is rather worrisome, since we are already at 385 ppm and are almost bound to reach 450 ppm before the level stabilizes, even if we get very serious very soon about cutting our greenhouse-gas emissions.

We cannot afford to stay at 450 ppm for very long: there is a grace period of only a few decades before the consequent warming in the climate leads to irreversible changes, including the eventual melting of all the world's ice. The various agricultural and forestry changes that Flannery discusses will be of great use in getting that extra carbon dioxide back out of the atmosphere in the long run — but it is quite a long run, almost certainly longer than the time available if that level of carbon dioxide is allowed to translate into an equivalent rise in temperature.

But that does not have to happen. The dirty little secret is that we know of several techniques for keeping the global average temperature down, even though the carbon-dioxide concentration implies a hotter planet. These are not long-term solutions, because they do nothing to slow ocean acidification and do not necessarily produce cooling in the parts of the planet that need it most, but as stopgap measures to keep us from breaking through the three- to four-degree barrier they are probably going to be indispensable for a while. They may be the only way that we can win extra time to work on getting our emissions down without breaching the limits and hitting runaway warming.

Flannery knows this, and even makes reference to one possible geo-engineering technique — Crutzen's sulfur-in-the-stratosphere proposal — but I do not think he gives the subject the prominence it deserves. These are techniques that may be crucial to our chances of getting through this without a calamity of global proportions, and they need to be researched and tested aggressively now. We may find that we need them quite soon.

GWYNNE DYER is a freelance journalist, columnist, broadcaster, and lecturer on international affairs. He is the author of several books, including *War*; *Future: Tense*; *The Mess They Made*; and *Climate Wars*.

Alanna Mitchell

I was jet-lagged in Brisbane the first time I read Tim Flannery's splendid essay and originally thought it was about Australians and Australia. But on reflection, I realized that it is an invocation to all humanity and that Flannery's homeland is an elegant metaphor for the planet as a whole.

It is a single system, after all, and we humans are a single, messy species. We are connected with each other and with everything else on Earth, despite all that our powerful tribalistic stories tell us to the contrary. The very plasma that courses through our veins has the same magical chemistry as the ocean's planet-blood that gave birth to life.

As his essay shows, few know the scope of this ineluctable connectedness better than Flannery. Few have sensed more deeply what will happen if the system lurches into a new, post-human mode. And Flannery is alone in his uncanny ability to explain what we can do to try to prevent that. All this has made him, in my view, the world's foremost oracle.

But as I sat in hotel room after hotel room through September 2008, shifting to Sydney, Melbourne, Adelaide, and finally Hobart, watching the world's financial institutions turn to puddles on the global trading floor, watching the American government try in vain to shape them up again, and watching the panic escalate, I admit to sheer marvel at Flannery's sense of timing.

Here, written in trillions of dollars and billions of lives, were the makings of just such a system shift as Flannery has been telling us about. The meltdown is a searing example of the logarithmic system change that scientists predict for the biological world, played out in the language of money instead.

It wasn't neat and orderly. It was bewilderingly chaotic. It just kept going, throwing new curves at dazed observers. Here Lehman Brothers fell. There, the banking industry as a whole threatened to topple, leading to unknown consequences. Stocks lost trillions. Brokers, once great strutters, began to slouch and then to slither. That wash of money that investment bankers had seemed to handle so deftly had somehow taken on a mind of its own. What started out affecting millions grew to billions, and then to trillions. Suddenly, no one was immune. Terrified national governments conjured up hundreds of billions of dollars and scores of billions of pounds and euros to shore up the tidal wave. Had that money not materialized, the financial system was poised to morph into a different beast altogether.

This is the definition of system switch, the same phenomenon Flannery foretells in the realm of Gaia if we do not—quickly—step in.

Time will tell if the financial intervention worked. The principle, though, is sound: once a system begins to tilt, it takes heroic efforts to convince it not to shift merrily, heedlessly, catastrophically into something wholly new. And then you have to figure out what the causes of the breakdown were and also the trends underpinning the cause, and fix them. (This is where Nicolas Sarkozy, the French president, has come in so handy lately, talking about what the capitalist financial order was *for* in the first place.)

A couple of larger points emerge from the muddle. Some of the public discourse has been about whether we should now just hunker down and forget about going green, about investing in the sorts of world-saving technologies that Flannery tells us about in his essay. I heard it the other day in Toronto at a conference on corporate social responsibility. Maybe, a fellow or two opined, we should pull in our horns, stick to the old-fashioned assessment of risk, focus on the short term until the heavy weather clears. In other words, save a system just so it can crash again. Because, as is clear, the financial system nearly went under because it couldn't keep going under the same circumstances that had brought it to its knees. Something had to change. It wasn't sustainable. All that profligate profit was based on something that wasn't there.

So going brown, financially speaking, is precisely the wrong answer. This is, as Matthew Kiernan explained the other day, a "teaching moment." He is chief executive of Canada's Innovest Strategic Value Advisers and one of the braver souls at the Toronto conference. To him, the crash is a "trillion-dollar advertorial" for expanding the narrow, traditional definition of financial risk. In his view, assessing risk—and therefore opportunity—must include looking at elements such as environment, human rights, politics, labor markets, and even health and safety practices. "The current paradigm is broken," he says. "What better moment?" The lesson is to keep going down the green financial road.

I think we could go farther. What if the world's trillions in investments could be used to produce profit *as well as* social and environmental good? What if they were seen as two sides of the same coin and it were acknowledged that one was absent without the other? The planet's carbon-dioxide concentrations could be lowered swiftly and efficiently. It's not as crazy as it sounds, and very smart financial people have been looking at this seriously, now that carbon has become a global commodity with a value. We are all one system. All of this is interconnected. Finance and biology are, in thought and deed, parts of a whole.

Flannery's timing is impeccable. Not only did his essay come out just at the moment when, for the first time in generations, all assumptions seemed to be collapsing; not only was the biological meltdown he refers to brought to life in the financial markets before our very eyes; not only are his Australian suggestions for breaking the untenable cycle a poignant recipe book for other countries; but the essay also, in its rich faith in humanity's ability to cope with the exigencies upon us, offers up the gift of poetry, and with it, the power of hope.

ALANNA MITCHELL is the author of *Seasick: The Hidden Ecological Crisis of the Global Ocean* (2008) and *Dancing at the Dead Sea: Journey to the Heart of Environmental Crisis* (2005). She lives in Toronto.

REPLY

Reply by Tim Flannery

I am immensely grateful to the respondents for taking the time to craft such thoughtful comments on my tract. Some, such as Bill McKibben and Fred Krupp, have far longer experience in the climate change arena than I, and I am heartened by their optimism. Others, such as Sir Richard Branson, are among the busiest of people working at the industrial "coal face," trying to make a difference in such difficult areas as aviation. All these respondents have important things to say that amplify and round out my basic message. I do, however, have a few quibbles with some ideas put forward.

Krupp and Goldmark's criticism of the "clean development mechanism" (CDM) of the European carbon trading scheme under the Kyoto Protocol is, I feel, a little too harsh. Certainly it has been flawed, but not to the point of having a "negligible" impact. But more important, with Kyoto coming to an end in 2012 and (it is hoped) a new treaty commencing in

2013, there is an opportunity to revise the CDM, thereby making it more effective. Although I agree that a broader approach would be even better, it would be a pity if mechanisms like the CDM were abandoned altogether; the investments they encourage in developing countries are potentially powerful tools for reducing emissions.

The most challenging response is without doubt that of Peter Singer. He has many perceptive things to say and is quite correct in his assessment of the impacts of methane and nitrous oxide over a twenty-year time horizon as opposed to one hundred years. He is also correct to state that the livestock sector is a significant contributor to these emissions. I disagree with him, however, over how we deal with the problem. Peter says that we should simply stop eating beef. It's a deceptively simple solution, because it ignores the fact that the production of some beef results in far less greenhouse gas emissions than the production of other beef, and that some forms of vegetarian foods result in far higher emissions than other kinds. For example, intensive use of agricultural soils to grow crops has, over the past century, resulted in the loss of two-thirds of their soil carbon—about sixty-six gigatonnes in all. And in the beef sector, it's been found that smaller breeds of cattle produce 25 percent less methane than standard breeds, and that the overall management of the herd has an enormous impact on the overall greenhouse gas balance of the business.

I do not wish to downplay the importance of Singer's contribution, however. His emphasis in trying to bring some rigor to our analysis of how sustainably various kinds of meat are produced is vitally important. He quite rightly notes that the majority of cattle produced in developed countries are processed through feedlots, and there is no doubt that this practice produces a voluminous waste stream of greenhouse gases. Such practices need to be targeted—and those who practice them made accountable—as we address climate change, and Singer will doubtless play an important role in this. There are, however, more sustainable models of livestock management, including "holistic management" of rangelands, that result in a drawdown of CO_2 into rangelands soils. Singer has not, in my opinion, given sufficient weight to the overall greenhouse accounting of such options. And he is wrong about fire and rangelands. Good livestock management results in healthier soils and greater carbon sequestration than occurs on unmanaged lands.

Furthermore, I remain unconvinced by Singer's analysis of the overall greenhouse gas balance on Polyface Farm. As mentioned above, methane emissions and nitrous oxide emissions from cattle vary enormously with husbandry, and trying to account for Polyface's emissions from data elsewhere is not likely to shed much light. In order to truly understand emissions from a place like Polyface Farm, a local audit is required. One benefit of Singer's focus is that such audits are likely to become more common in the future, and these will aid farmers in reducing their greenhouse gas impact.

I must say, however, that I'm dismayed by Singer's final words: "There are no excuses left for eating beef." This riposte entirely ignores my suggestion that we adopt a "sustainabilitarian" diet—one that is good both for the human body and for the planet. We would be best served in such matters if we treat the meat, vegetable, coal, and aviation industries similarly in our demands that they reduce their emissions. As I have argued in my essay, this means demanding that

the coal industry shift quickly to clean coal technologies. For the growers of vegetables and grain it means
growing as locally as possible and using demonstra-
tively sustainable production methods. For meat and
aviation it means a shift, as soon as practicable, to an
overall zero or negative emissions profile.

The last word on this issue should, I feel, go to Sir
Richard, for he is in the position of being responsible
for such a transition. "What does an economy that uses
only the resources of a single planet look like?" he asks
us. There's no doubt that the livestock industry, like
airlines, will need to change profoundly over this century. Neither, however, should be put out of business
for ideological rather than sustainabilitarian reasons.

Now or Never is printed on Ancient Forest Friendly paper, made with 100% post-consumer waste.